Martingales and Financial Mathematics in Discrete Time

Series Editor
Nikolaos Limnios

Martingales and Financial Mathematics in Discrete Time

Benoîte de Saporta
Mounir Zili

ISTE

WILEY

First published 2021 in Great Britain and the United States by ISTE Ltd and John Wiley & Sons, Inc.

ISTE Ltd
27-37 St George's Road
London SW19 4EU
UK

www.iste.co.uk

John Wiley & Sons, Inc.
111 River Street
Hoboken, NJ 07030
USA

www.wiley.com

Library of Congress Control Number: 2021944040

British Library Cataloguing-in-Publication Data
A CIP record for this book is available from the British Library
ISBN 978-1-78630-669-2

MIX
Paper from responsible sources
FSC® C013604

Contents

Preface ix

Introduction xi

Chapter 1. Elementary Probabilities and an Introduction to Stochastic Processes 1

1.1. Measures and σ-algebras 1
1.2. Probability elements 5
1.2.1. Probabilities 5
1.2.2. Random variables 8
1.2.3. σ-algebra generated by a random variable 12
1.2.4. Random vectors 13
1.2.5. Convergence of sequences of random variables 15
1.3. Stochastic processes 16
1.4. Exercises 19

Chapter 2. Conditional Expectation 21

2.1. Conditional probability with respect to an event 21
2.2. Conditional expectation 24
2.2.1. Definitions 24
2.2.2. Properties of conditional expectation 28
2.3. Geometric interpretation 37
2.4. Conditional expectation and independence 38
2.5. Exercises 41

Chapter 3. Random Walks 45

3.1. Trajectories of the random walk 45
3.1.1. Definition 45
3.1.2. Graphical representation 47
3.1.3. Reflection principle 50
3.2. Asymptotic behavior 52
3.2.1. The Markov property and stationarity property 52
3.2.2. First return to 0 55
3.3. The Gambler's ruin 58
3.4. Exercises 60

Chapter 4. Martingales 63

4.1. Definition 63
4.2. Martingale transform 66
4.3. The Doob decomposition 67
4.4. Stopping time 69
4.5. Stopped martingales 71
4.6. Exercises 75

Chapter 5. Financial Markets 81

5.1. Financial assets 82
5.2. Investment strategies 82
5.3. Arbitrage 84
5.4. The Cox, Ross and Rubinstein model 86
5.5. Exercises 88
5.6. Practical work 90
5.6.1. Simulation of trajectories 90
5.6.2. Portfolio optimization 90
5.6.3. Portfolio optimization with withdrawal 91

Chapter 6. European Options 95

6.1. Definition 95
6.2. Complete markets 96
6.3. Valuation and hedging 97
6.4. Cox, Ross and Rubinstein model 98
6.4.1. Completeness 98
6.4.2. Value of European options 101
6.4.3. Hedging European options 103
6.5. Exercises 104
6.6. Practical work: Simulating the value of a call option 106

Chapter 7. American Options 107

7.1. Definition 107
7.2. Optimal stopping 109
7.2.1. Snell envelope and stopping time 109
7.2.2. Construction of optimal stopping times 111
7.3. Application to American options 114
7.4. The Cox, Ross and Rubinstein model 115
7.4.1. Value of American options 115
7.4.2. Hedging American options 116
7.5. Exercises 116
7.6. Practical work 117

Chapter 8. Solutions to Exercises and Practical Work 119

8.1. Solutions to exercises in Chapter 1 119
8.1.1. Exercise 1.1 119
8.1.2. Exercise 1.2 119
8.1.3. Exercise 1.3 120
8.1.4. Exercise 1.4 120
8.1.5. Exercise 1.5 121
8.1.6. Exercise 1.6 122
8.1.7. Exercise 1.7 123
8.1.8. Exercise 1.8 125
8.2. Solutions to exercises in Chapter 2 127
8.2.1. Exercise 2.1 127
8.2.2. Exercise 2.2 129
8.2.3. Exercise 2.3 131
8.2.4. Exercise 2.4 133
8.2.5. Exercise 2.5 133
8.2.6. Exercise 2.6 135
8.2.7. Exercise 2.7 135
8.2.8. Exercise 2.8 136
8.2.9. Exercise 2.9 138
8.2.10. Exercise 2.10 138
8.2.11. Exercise 2.11 141
8.2.12. Exercise 2.12 142
8.3. Solutions to exercises in Chapter 3 143
8.3.1. Exercise 3.1 143
8.3.2. Exercise 3.2 144
8.3.3. Exercise 3.3 144
8.3.4. Exercise 3.4 146
8.3.5. Exercise 3.5 147
8.3.6. Exercise 3.6 148
8.3.7. Exercise 3.7 148
8.3.8. Exercise 3.8 149

8.4. Solutions to exercises in Chapter 4 151
8.4.1. Exercise 4.1 151
8.4.2. Exercise 4.2 151
8.4.3. Exercise 4.3 152
8.4.4. Exercise 4.4 153
8.4.5. Exercise 4.5 154
8.4.6. Exercise 4.6 157
8.4.7. Exercise 4.7 160
8.4.8. Exercise 4.8 163
8.4.9. Exercise 4.9 164
8.4.10. Exercise 4.10 166
8.4.11. Exercise 4.11 167
8.5. Solutions to exercises in Chapter 5 170
8.5.1. Exercise 5.1 170
8.5.2. Exercise 5.2 171
8.5.3. Exercise 5.3 173
8.5.4. Exercise 5.4 174
8.5.5. Exercise 5.5 174
8.6. Solutions to the practical exercises in Chapter 5 175
8.6.1. Practical exercise 5.6.1 175
8.6.2. Practical exercise 5.6.2 176
8.6.3. Practical exercise 5.6.3 183
8.7. Solutions to exercises in Chapter 6 189
8.7.1. Exercise 6.1 189
8.7.2. Exercise 6.2 190
8.7.3. Exercise 6.3 191
8.7.4. Exercise 6.4 193
8.8. Solution to the practical exercise in Chapter 6 (section 6.6) 193
8.9. Solution to exercises in Chapter 7 195
8.9.1. Exercise 7.1 195
8.9.2. Exercise 7.2 196
8.9.3. Exercise 7.3 197
8.10. Solution to the practical exercise in Chapter 7 (section 7.6) 200

References 205

Index 207

Preface

The authors owe a lot to many people and institutions who have contributed in diverse ways to this book. Most of the material presented in this book was used in various courses at different levels and for different audiences, being constantly amended and improved based on the feedback received from colleagues and students.

About half of these chapters were initially designed for students studying *Financial Engineering* as part of their MIMSE master's program (Engineering Mathematics, Statistics and Economics), created in 2007 and jointly offered by the Universities of Bordeaux 1, Bordeaux 2 and Montesquieu Bordeaux IV, which then merged to form the present-day University of Bordeaux. The students who took this course were awarded their bachelor's either in economics or mathematics. Some parts of this course benefited from the experience of and collaboration with other professors such as François Dufour and Christine Marois.

Some chapters in this book are also the results of notes created by a group of teachers on stochastic processes, which were given to students, from 2013 onward, undergoing the course *Mathematics and applications* as part of their Research Masters under the Faculty of Science in the University of Monastir. In the framework of this stream, and following the initiative of Ali Gannoun and Leila Ben Abdelghani Bouraoui, the course material at Montpellier was also adapted in 2017 for another unit that taught Financial Mathematics in the Faculty of Science at Monastir.

Other chapters were first designed to be a part of the Stochastic Processes course in the *Biostatistics* and *Mathematics of Information and Decision-Making* streams in the Masters in Mathematics offered by the University of Montpellier.

The authors extend their warmest gratitude to all the people and institutions who have contributed to the creation of this book.

The authors wish to dedicate this work to Leila Ben Abdelghani Bouraoui and Ali Gannoun.

Benoîte DE SAPORTA
Mounir ZILI
August 2021

Introduction

Ever since the work of Black-Scholes-Merton [BLA 73] in 1973, the design, analysis and development of complex financial products and services have required not only an ever-greater understanding of financial theories but also a mastery of probability theory and stochastic processes.

This book introduces basic concepts of this theory, especially that of discrete-time martingales. It shows how this concept can be applied to the pricing and hedging of derivatives in financial markets. There are many texts in the existing literature that focus on financial mathematics and the evaluation of options, for example [LAM 97, SHR 04, VIN 04]. The majority of these texts consider only continuous-time models and make extensive use of Itô calculus. A few rare books such as [SHR 03] introduce discrete-time studies, but this is also done to prepare readers to understand continuous-time financial markets. The most unique feature of this book is that it is entirely devoted to discrete time and provides a detailed introduction to the construction of the rigorous mathematical tools required for the evaluation of options in financial markets. The only pre-requisite for this book is a basic understanding of probability. Several theoretical and numerical aspects are studied in this book, explored through multiple examples and exercises for which complete solutions are provided.

To the best of the authors' knowledge, this book is the first reference to include both mathematical teaching with multiple exercises that have detailed solutions, focusing on discrete martingales and their application to financial markets, and practical work with solutions, using the software R. Special attention is paid to the Cox, Ross and Rubinstein model in discrete time.

This book is meant for students at the master's or doctoral level who are specializing in applied mathematics or finance as well as teachers, researchers in the field of economics or actuarial science, professionals working in the various finance sectors, or any other person who may be interested in a rigorous and accessible mathematical construction of the tools and concepts used in financial mathematics or the application of the martingale theory in finance. Finally, the practical work on optimal portfolio management in Chapter 5, as well as the study of American options in Chapter 7, may serve as an elementary introduction to stochastic control problems.

This book is arranged as follows: Chapter 1 reviews the basic concepts and notations in probability and random variables, especially discrete variables, and also provides an introduction to the concept of stochastic processes. Chapter 2 is dedicated to the introduction and study of the concept of conditional expectation, a key concept in the definition of martingales and the computation of financial options. Chapter 3 aims to introduce an interesting example of stochastic process, namely a simple symmetric random walk, with minimal formalism. This chapter may be read as a standalone chapter. Chapter 4 defines and characterizes the concept of a martingale in discrete time and studies certain properties.

From Chapter 5 onward, we focus on financial mathematics, strictly speaking. We define essential financial vocabulary such as the concept of a financial asset, investment strategy and the concept of arbitrage. We also begin to establish a link with martingales. We also introduce a typical example of a discrete financial market: the Cox, Ross and Rubinstein binomial model, which acts as a guiding thread through all the following chapters. In particular, the question of Optimal Portfolio Management in this model is studied through guided work. This is a discretized version of the famous Merton problem, originally posed in the continuous Black and Scholes [MER 69] model. Chapter 6 introduces and studies the first large and specific category of conditional assets, i.e. European options. This is the simplest example of a financial asset which may be exercised subject to the realization of a certain condition. We examine, in detail, the question of the pricing and hedging of these options in the general case of a discrete financial market and then in the specific case of the Cox, Ross and Rubinstein model. Chapter 7 is dedicated to the study of a slightly more complex family of conditional assets: American options. We once again undertake a detailed study of the question of the pricing and hedging of these options and connect them to the theory of optimal stopping, in the general case of a discrete financial market and then in the specific case of the Cox, Ross and Rubinstein model.

The detailed solution to all the exercises and practical work are given together at the end of the book in Chapter 8.

1

Elementary Probabilities and an Introduction to Stochastic Processes

This chapter reviews the basic concepts related to probability and random variables which will be useful for the rest of this text. For a more detailed explanation as well as demonstrations, the readers may refer to [BAR 07, DAC 82, FOA 03, OUV 08, OUV 09] in French and [BIL 12, CHU 01, DUR 10, KAL 02, SHI 00] in English. The readers who are already familiar with these concepts may proceed straight to section 1.3, which introduces the concept of stochastic processes.

This chapter begins with a brief summary of the concepts of a σ-algebra in section 1.1. These concepts will help in understanding the construction of the properties of conditional expectation in Chapter 2. We then study the chief definitions and properties of random variables and their distribution in section 1.2. There is an emphasis on discrete random variables as this entire book essentially studies discrete cases. Section 1.3 defines a stochastic process, which is the main subject studied in this book. Finally, there are exercises in handling these different concepts in section 1.4. The solutions are given in Chapter 8.

Throughout the rest of the text, Ω is a non-empty set and $\mathcal{P}(\Omega)$ denotes the set of the subsets of Ω :

$$\mathcal{P}(\Omega) = \{A; A \subset \Omega\}.$$

The set Ω is called the **universe** or the **fundamental set**. In practice, the set Ω contains all the possible outcomes of a random experiment.

1.1. Measures and σ-algebras

Let us start by reviewing the concept of a σ-algebra.

DEFINITION 1.1.– *A subset $\mathcal{F}$ of $\mathcal{P}(\Omega)$ is a σ-algebra over Ω if*

1) $\Omega \in \mathcal{F}$;

2) $\mathcal{F}$ is stable by complementarity: for any $A \in \mathcal{F}$, we have $A^c \in \mathcal{F}$, where A^c denotes the complement of A in Ω: $A^c = \Omega \backslash A$;

3) $\mathcal{F}$ is stable under a countable union: for any sequence of elements $(A_n)_{n\in\mathbb{N}}$ of $\mathcal{F}$, we have $\bigcup_{n\in\mathbb{N}} A_n \in \mathcal{F}$.

*Elements of a σ-algebra are called **events**.*

EXAMPLE 1.1.– *The set $\mathcal{G} = \{\emptyset, \Omega\}$ is a σ-algebra and is also the smallest σ-algebra over Ω; it is called the **trivial σ-algebra**. Indeed, $\mathcal{G}$ is in fact a σ-algebra since $\Omega \in \mathcal{G}$, $\emptyset^c = \Omega \in \mathcal{G}$, $\Omega^c = \emptyset \in \mathcal{G}$, $\Omega \cup \emptyset = \Omega \in \mathcal{G}$ and by creating unions of $\emptyset$ and Ω we always obtain $\emptyset \in \mathcal{G}$ or $\Omega \in \mathcal{G}$. Further, for any other σ-algebra $\mathcal{F}$, we clearly have $\mathcal{G} \subset \mathcal{F}$.* ◇

EXAMPLE 1.2.– *The set $\mathcal{P}(\Omega)$ is the largest σ-algebra over Ω; it is called the **largest σ-algebra.** Indeed, by construction, $\mathcal{P}(\Omega)$ contains all the subsets of Ω, and thus it contains in particular Ω and it is stable by complementarity and under countable unions. In addition, any other σ-algebra $\mathcal{F}$ over Ω is clearly included in $\mathcal{P}(\Omega)$.* ◇

DEFINITION 1.2.– *Let Ω be a non-empty set and $\mathcal{F}$ be a σ-algebra over Ω. The couple $(\Omega, \mathcal{F})$ is called a **probabilizable space**.*

Among the elementary properties of σ-algebra, we can cite stability through any intersection (countable or not).

PROPOSITION 1.1.– *Any intersection of σ-algebras over a set Ω is a σ-algebra over Ω.*

PROOF.– Let $(\mathcal{F}_i)_{i\in I}$ be any family of σ-algebra indexed by a non-empty set I. Thus,

– first of all, for any i, $\Omega \in \mathcal{F}_i$, thus $\Omega \in \cap_{i\in I}\mathcal{F}_i$;

– secondly, if $A \in \cap_{i\in I}\mathcal{F}_i$, then for any i, $A \in \mathcal{F}_i$. As these are σ-algebras, we have that for any i, $A^c \in \mathcal{F}_i$, thus $A^c \in \cap_{i\in I}\mathcal{F}_i$;

– finally, if for any $n \in \mathbb{N}$, $A_n \in \cap_{i\in I}\mathcal{F}_i$, then for any i, n, $A_n \in \mathcal{F}_i$. As these are σ-algebras, we have that for any i, $\cup_{n\in\mathbb{N}}A_n \in \mathcal{F}_i$, thus

$$\cup_{n\in\mathbb{N}}A_n \in \cap_{i\in I}\mathcal{F}_i.$$ □

It is generally difficult to make explicit all the events in a σ-algebra. We often describe it using generating events.

DEFINITION 1.3.– *Let $\mathcal{E}$ be a subset of $\mathcal{P}(\Omega)$. The σ-algebra $\sigma(\mathcal{E})$ **generated** by $\mathcal{E}$ is the intersection of all σ-algebras containing $\mathcal{E}$. It is the smallest σ-algebra containing $\mathcal{E}$. $\mathcal{E}$ is called the **generating system** of the σ-algebra $\sigma(\mathcal{E})$.*

It can be seen that $\sigma(\mathcal{E})$ is indeed a σ-algebra, being an intersection of σ-algebras.

EXAMPLE 1.3.– *If $A \subset \Omega$, then, $\sigma(A) = \{\emptyset, \Omega, A, A^c\}$ is the smallest σ-algebra Ω containing A.* ◇

EXAMPLE 1.4.– *If Ω is a topological space, the σ-algebra generated by the open sets of Ω is called the Borel **σ-algebra** of Ω. A **Borel set** is a set belonging to the Borel σ-algebra. On $\mathbb{R}$, $\mathcal{B}(\mathbb{R})$ generally denotes the σ-algebra of Borel sets. It must be recalled that this is also the σ-algebra generated by the intervals, or by the intervals of the form $]-\infty, x]$, $x \in \mathbb{R}$. Thus, there is no unicity of the generating system.* ◇

We will now recall the concept of the product σ-algebra.

DEFINITION 1.4.– *Let $(E_i, \mathcal{F}_i)_{i \in \mathbb{N}}$ be a sequence of measurable spaces.*

– *Let $n \in \mathbb{N}$. The σ-algebra defined over $\prod_{i=0}^{n} E_i$ and generated by*

$$\{A_0 \times ... \times A_n; A_0 \in \mathcal{F}_0, ..., A_n \in \mathcal{F}_n\}$$

is denoted by $\mathcal{F}_0 \otimes ... \otimes \mathcal{F}_n$, and it is called the product σ-algebra over $\prod_{i=0}^{n} E_i$. We have, in particular,

$$\mathcal{F}_0 \otimes ... \otimes \mathcal{F}_n = \sigma\left(A_0 \times ... \times A_n; A_0 \in \mathcal{F}_0, ..., A_n \in \mathcal{F}_n\right).$$

In the specific case where $E_0 = ... = E_n = E$ and $\mathcal{F}_0 = ... = \mathcal{F}_n = \mathcal{F}$, we also write

$$\prod_{i=0}^{n} E_i = E^{n+1} \text{ and } \mathcal{F}_0 \otimes ... \otimes \mathcal{F}_n = \mathcal{F}^{\otimes n+1}.$$

– *We use $\otimes_{i \in \mathbb{N}} \mathcal{F}_i$ to denote the σ-algebra over the countable product space $\prod_{i \in \mathbb{N}} E_i$, generated by the sets of the form $\prod_{i \in \mathbb{N}} A_i$, where $A_i \in \mathcal{F}_i$ and $A_i = E_i$ except for a finite number of indices i. In the specific case where, for any $i \in \mathbb{N}$, $E_i = E$ and $\mathcal{F}_i = \mathcal{F}$, the product space $\prod_{i \in \mathbb{N}} E_i$ is denoted by $E^{\mathbb{N}}$, and the σ-algebra $\otimes_{i \in \mathbb{N}} \mathcal{F}_i$ is denoted by $\mathcal{F}^{\otimes \mathbb{N}}$.*

Finally, let us review the concepts of measurability and measure.

DEFINITION 1.5.– *Let Ω be non-empty set and $\mathcal{F}$ be a σ-algebra on Ω.*

*– A **measure** over a probabilizable space $(\Omega, \mathcal{F})$ is defined as any mapping μ defined over $\mathcal{F}$, with values in $[0, +\infty] = \mathbb{R}_+ \cup \{+\infty\}$, such that $\mu(\emptyset) = 0$ and for any family $(A_i)_{i \in \mathbb{N}}$ of pairwise disjoint elements of $\mathcal{F}$, we have the property of σ-additivity:*

$$\mu\left(\bigcup_{i=0}^{+\infty} A_i\right) = \sum_{i=0}^{+\infty} \mu(A_i).$$

*– A measure μ over a probabilizable space $(\Omega, \mathcal{F})$ is said to be **finite**, or have **finite total mass**, if $\mu(\Omega) < \infty$.*

*– If μ is a **measure** over a probabilizable space $(\Omega, \mathcal{F})$, then the triplet $(\Omega, \mathcal{F}, \mu)$ is called a **measured space**.*

DEFINITION 1.6.– *Let $(\Omega, \mathcal{F})$ and $(E, \mathcal{E})$ be two probabilizable spaces. A mapping X, defined over Ω taking values in E, is said to be $(\mathcal{F}, \mathcal{E})$-**measurable**, or just measurable, if there is no ambiguity regarding the reference σ-algebras, if*

$$\forall B \in \mathcal{E}, \;\; X^{-1}(B) \in \mathcal{F}.$$

In practice, when $E \subset \mathbb{R}$, we set $\mathcal{E} = \mathcal{B}(E)$ the set of Borel subsets of E, that is, the set of subsets of E. We can simply say that X is $\mathcal{F}$-measurable. When, in addition, we manipulate a single σ-algebra $\mathcal{F}$ over Ω, it can be simply said that X is measurable. If we work with several σ-algebras over Ω, the concerned σ-algebra must always be specified: X is $\mathcal{F}$-measurable.

EXAMPLE 1.5.– *If $(\Omega, \mathcal{F})$ is a measurable space and $A \in \mathcal{F}$, then the indicator function*

$$\mathbb{1}_A : x \longmapsto \begin{cases} 1 \; if \; x \in A \\ 0 \; if \; x \notin A \end{cases}$$

is $\mathcal{F}$-measurable. Indeed, for any Borel set B in $\mathbb{R}$, we have

$$(\mathbb{1}_A)^{-1}(B) = \begin{cases} \Omega \; if \; \{0,1\} \subset B \\ A \; if \; 1 \in B \; and \; 0 \notin B \\ A^c \; if \; 0 \in B \; and \; 1 \notin B \\ \emptyset \; if \; 0 \notin B \; and \; 1 \notin B \end{cases}$$

Thus, in all cases, we do have $(\mathbb{1}_A \in B) \in \mathcal{F}$. ◇

EXAMPLE 1.6.– *The composition of two measurable functions is measurable. Indeed, if $(\Omega, \mathcal{F})$, $(E, \mathcal{E})$ and $(G, \mathcal{G})$ are three probabilizable spaces, $f : \Omega \longmapsto E$ and*

$g : E \longmapsto G$ are two $(\mathcal{F}, \mathcal{E})$ and $(\mathcal{E}, \mathcal{G})$-measurable mappings, respectively, then for any $B \in \mathcal{G}$, $g^{-1}(B) \in \mathcal{E}$ and consequently,

$$(g \circ f)^{-1}(B) = f^{-1}(g^{-1}(B)) \in \mathcal{F}.$$

Thus, the composition $g \circ f$ is indeed measurable on $(\Omega, \mathcal{F})$ in $(G, \mathcal{G})$. ◇

1.2. Probability elements

We will now review the concept of a probability measure or probability distribution, and the concept of random variable, as well as the chief properties of these concepts.

1.2.1. *Probabilities*

A probability measure or probability distribution is a finite measure whose total mass is equal to 1.

DEFINITION 1.7.– *A **probability** or **probability measure**, or **law of probability** or **distribution** over a probability space $(\Omega, \mathcal{F})$ is a measure with a total mass equal to* 1. *In other words, a probability over $(\Omega, \mathcal{F})$ is a mapping $\mathbb{P} : \mathcal{F} \longrightarrow \mathbb{R}$ such that*

– *for any $A \in \mathcal{F}$, $\mathbb{P}(A) \geq 0$,*

– $\mathbb{P}(\Omega) = 1$,

– *for any sequence of pairwise disjoint events in $\mathcal{F}$, denoted by $(A_n)_{n \in \mathbb{N}}$, we have*

$$\mathbb{P}\left(\bigcup_{n=0}^{+\infty} A_n\right) = \sum_{n=0}^{+\infty} \mathbb{P}(A_n).$$

*The triplet $(\Omega, \mathcal{F}, \mathbb{P})$ is then called a **probability space**.*

EXAMPLE 1.7.– *Ω is endowed with the coarse σ-algebra $\mathcal{F} = \{\emptyset, \Omega\}$. Thus, the single probability measure on $(\Omega, \mathcal{F})$ is given by:*

$$\mathbb{P} : A \in \mathcal{F} \longmapsto \mathbb{P}(A) = \begin{cases} 1 \; if \; A = \Omega \\ 0 \; if \; A = \emptyset. \end{cases}$$ ◇

EXAMPLE 1.8.– *Let $\Omega = [0, 1]$ and $\mathcal{F} = \mathcal{B}([0, 1])$ be the Borel σ-algebra of $[0, 1]$. If λ denotes the Lebesgue measure, then the mapping:*

$$\mathbb{P} : A \in \mathcal{F} \longmapsto \lambda(A)$$

is a probability measure on $(\Omega, \mathcal{F})$. ◇

EXAMPLE 1.9.– *Let Ω be non-empty set such that $card(\Omega) < \infty$, where $card(\Omega)$ denotes the cardinal of Ω, that is, the number of elements in Ω. Consider the mapping $\mathbb{P}$ from $\mathcal{P}(\Omega)$ onto $[0,1]$ such that for every $A \in \mathcal{P}(\Omega)$, $\mathbb{P}(A) = \dfrac{card(A)}{card(\Omega)}$.*

The mapping $\mathbb{P}$ is then a probability on $(\Omega, \mathcal{P}(\Omega))$, said to be the uniform probability on Ω. ◇

We will only review those properties of a probability that will be useful for this book.

PROPOSITION 1.2.– *Let $(\Omega, \mathcal{F}, \mathbb{P})$ be a probability space and $(A_n)_{n\in\mathbb{N}}$ be a sequence of events in $\mathcal{F}$.*

– If $(A_n)_{n\in\mathbb{N}}$ is increasing (for the inclusion), then,

$$\mathbb{P}\left(\bigcup_{i=0}^{\infty} A_i\right) = \lim_{n\to+\infty} \mathbb{P}(A_n).$$

– If $(A_n)_{n\in\mathbb{N}}$ is decreasing (for the inclusion), then,

$$\mathbb{P}\left(\bigcap_{i=0}^{\infty} A_i\right) = \lim_{n\to+\infty} \mathbb{P}(A_n).$$

We will now review the concept of independent events and σ-algebras.

DEFINITION 1.8.– *Let $(\Omega, \mathcal{F}, \mathbb{P})$ be a probability space.*

– Two events, A and B, are **independent** *if $\mathbb{P}(A \cap B) = \mathbb{P}(A) \times \mathbb{P}(B)$.*

– A family of events $(A_i \in \mathcal{F}_i, i \in I)$ is said to be **mutually independent** *if for any finite family $J \subset I$, we have*

$$\mathbb{P}\left(\bigcap_{j\in J} A_j\right) = \prod_{j\in J} \mathbb{P}(A_j).$$

– Two σ-algebras $\mathcal{F}$ and $\mathcal{G}$ are independent if for any $A \in \mathcal{F}$ and $B \in \mathcal{G}$, A and B are independent.

– A family of sub-σ-algebra $\mathcal{F}_i \subset \mathcal{F}, i \in I$ is **mutually independent** *if any family of events $(A_i \in \mathcal{F}_i, i \in I)$ is mutually independent.*

EXAMPLE 1.10.– *We roll a six-faced die and write*

– A_1 *the event "the number obtained is even"; and*

– A_2 *the event "the number obtained is a multiple of* 3*"*.

The universe of possible outcomes is $\Omega = \{1, 2, 3, 4, 5, 6\}$ *which has a finite number of elements and as all its elements have the same chance of occurring, we can endow it with the uniform probability* $\mathbb{P} : A \in \mathcal{P}(\Omega) \longmapsto \dfrac{card(A)}{card(\Omega)}$. *Since*

$$A_1 = \{2, 4, 6\},\ A_2 = \{3, 6\} \text{ and } A_1 \cap A_2 = \{6\},$$

we have

$$\mathbb{P}(A_1) = \frac{1}{2},\ \mathbb{P}(A_2) = \frac{1}{3} \text{ and } \mathbb{P}(A_1 \cap A_2) = \frac{1}{6} = \mathbb{P}(A_1) \times \mathbb{P}(A_2).$$

Therefore, A_1 *and* A_2 *are two independent events.* ◇

EXAMPLE 1.11.– *A coin is tossed twice. The following events are considered:*

– A_1 *"Obtaining tails (T) on the first toss";*

– A_2 *"Obtaining heads (H) on the second toss"; and*

– A_3 *"Obtaining the same face on both tosses".*

The universe of possible outcomes is

$$\Omega = \{(T, H), (T, T), (H, T), (H, H)\}$$

which has four elements, and as all elements have the same chance of occurring, it can be endowed with uniform probability. Since

$$A_1 = \{(T, H), (T, T)\},$$
$$A_2 = \{(T, H), (H, H)\},$$
$$A_3 = \{(T, T), (H, H)\},$$

we have

$$\mathbb{P}(A_1) = \mathbb{P}(A_2) = \mathbb{P}(A_3) = \frac{1}{2},$$
$$\mathbb{P}(A_1 \cap A_2) = \mathbb{P}(A_2 \cap A_3) = \mathbb{P}(A_1 \cap A_3) = \frac{1}{4},$$

and $\mathbb{P}(A_1 \cap A_2 \cap A_3) = 0 \neq \mathbb{P}(A_1) \times \mathbb{P}(A_2) \times \mathbb{P}(A_3)$. *Thus, the events* A_1, A_2 *and* A_3 *are pairwise independent, but are not mutually independent. Unless specified, the notion of independence by default always signifies mutual independence and not pairwise independence.* ◇

1.2.2. *Random variables*

Let us now recall the definition of a generic random variable, and then the specific case of discrete random variables.

DEFINITION 1.9.– *Let $(\Omega, \mathcal{F}, \mathbb{P})$ be a probabilizable space and $(E, \mathcal{E})$ be a measurable space. A* ***random variable*** *on the probability space $(\Omega, \mathcal{F}, \mathbb{P})$ taking values in the measurable space $(E, \mathcal{E})$, is any mapping $X : \Omega \longrightarrow E$ such that, for any B in $\mathcal{E}$, $X^{-1}(B) \in \mathcal{F}$; in other words, $X : \Omega \longrightarrow E$ is a random variable if it is an $(\mathcal{F}, \mathcal{E})$-measurable mapping. We then write the event* "X belongs to B" *by*

$$X^{-1}(B) = \{\omega \in \Omega; X(\omega) \in B\} = (X \in B).$$

In the specific case where $E = \mathbb{R}$ and $\mathcal{E} = \mathcal{B}(\mathbb{R})$, the mapping X is called a ***real random variable****. If $E = \mathbb{R}^d$ with d ≥ 2, and $\mathcal{E} = \mathcal{B}(\mathbb{R}^d)$, the mapping X is said to be a* ***real random vector****.*

EXAMPLE 1.12.– *Let us return to the experiment where a six-sided die is rolled, where the set of possible outcomes is $\Omega = \{1, 2, 3, 4, 5, 6\}$, which is endowed with the uniform probability. Consider the following game:*

– *if the result is even, you win* 10 €*;*

– *if the result is odd, you win* 20 €*.*

This game can be modeled using the random variable $X : \Omega \longmapsto \{10, 20\}$, defined by:

$$X(\omega) = \begin{cases} 10 \text{ if } \omega \in \{2, 4, 6\} \\ 20 \text{ if } \omega \in \{1, 3, 5\}. \end{cases}$$

This mapping is a random variable, since for any $B \in \mathcal{P}(\{10, 20\})$, we have

$$(X \in B) = X^{-1}(B) = \begin{cases} \{2, 4, 6\} \text{ if } B = \{10\} \\ \{1, 3, 5\} \text{ if } B = \{20\} \\ \Omega \text{ if } B = \{10, 20\} \\ \emptyset \text{ if } B = \emptyset. \end{cases}$$

and all these events are in $\mathcal{P}(\Omega)$. ◇

DEFINITION 1.10.– *The* ***distribution*** *of a random variable X defined on $(\Omega, \mathcal{F}, \mathbb{P})$ taking values in $(E, \mathcal{E})$ is the mapping $\mathbb{P}_X : \mathcal{E} \longrightarrow [0, 1]$ such that, for any$B \in \mathcal{E}$,*

$$\mathbb{P}_X(B) = \mathbb{P}(X \in B).$$

The distribution of X is a probability distribution on $(E, \mathcal{E})$; it is also called the ***image distribution*** *of $\mathbb{P}$ by X.*

DEFINITION 1.11.– *A random real variable is* ***discrete*** *if $X(\Omega)$ is at most countable. In other words, if $X(\Omega) = \{x_i, i \in I\}$, where $I \subset \mathbb{N}$. In this case, the probability distribution of X is characterized by the family*

$$(\mathbb{P}(X = x_i))_{i \in I}\,.$$

EXAMPLE 1.13.– ***Uniform distribution**: Let $N \in \mathbb{N}^\star$ and $\{x_1, ..., x_N\} \subset \mathbb{R}$. Let X be a random variable on $(\Omega, \mathcal{F}, \mathbb{P})$ such that $X(\Omega) = \{x_1, ..., x_N\}$ and for any $i \in \{1, ..., N\}$,*

$$\mathbb{P}(X = x_i) = \frac{1}{N}.$$

It is then said that X follows a uniform distribution on $\{x_1, ..., x_N\}$. ◇

EXAMPLE 1.14.– ***The Bernoulli distribution**: Let $p \in [0, 1]$. Let X be a random variable on $(\Omega, \mathcal{F}, \mathbb{P})$ such that $X(\Omega) = \{0, 1\}$ and*

$$\mathbb{P}(X = 1) = p \;\; \textit{and} \;\; \mathbb{P}(X = 0) = 1 - p.$$

It is then said that X follows a Bernoulli distribution with parameter p, and we write $X \sim \mathcal{B}(p)$.

The Bernoulli distribution models random experiments with two possible outcomes: success, with probability p, and failure, with probability $1 - p$. This is the case in the following game. A coin is tossed N times. This experiment is modeled by $\Omega = \{T, H\}^N$, endowed with the σ-algebra of its subsets and the uniform distribution. For $1 \leq n \leq N$, the mappings X_n from Ω onto $\mathbb{R}$ are considered, defined by

$$X_n(\omega_1, \omega_2, \ldots, \omega_n, \ldots, \omega_N) = \mathbb{1}_{\{T\}}(\omega_n),$$

the number of tails at the nth toss. Thus, X_n, $1 \leq n \leq N$, are random real variables in the Bernoulli distribution with parameter $1/2$ if the coin is balanced. ◇

EXAMPLE 1.15.– ***Binomial distribution**: Let $p \in [0, 1]$, $N \in \mathbb{N}^\star$ and X be a random variable on $(\Omega, \mathcal{F}, \mathbb{P})$ such that $X(\Omega) = \{0, 1, ..., N\}$ and for any $k \in \{0, 1, ..., N\}$,*

$$\mathbb{P}(X = k) = C_N^k p^k (1 - p)^{N-k}.$$

It is then said that X follows a binomial distribution with parameters N and p, and we write $X \sim \mathcal{B}(N, p)$.

If the Bernoulli experiment with probability of success p is repeated N times, independently, then the binomial distribution is the distribution of the random *variable containing the number of successes at the end of the N repetitions of the experiment.* ♢

EXAMPLE 1.16.– ***Hypergeometric distribution**: Let n and N be two integers such that $n \leq N$, $p \in]0,1[$ such that $pN \in \mathbb{N}$, and let X be a random variable on $(\Omega, \mathcal{F}, \mathbb{P})$ such that*

$$X(\Omega) = [\max(0; n - N(1-p)), \min(n, Np)] \cap \mathbb{N}$$

and for any $k \in X(\Omega)$,

$$\mathbb{P}(X = k) = \frac{C_{Np}^{k} C_{N(1-p)}^{n-k}}{C_N^n}.$$

X is then said to follow a hypergeometric distribution with parameters N, n and p, and we write $X \sim \mathcal{H}(N, n, p)$.

If we consider an urn containing N indistinguishable balls, k red balls and $N - k$ white balls, with $k \in \{1, ...N-1\}$, and if we simultaneously draw n balls, then the random variable X, equal to the number of red balls obtained, follows a hypergeometric distribution with parameters N, n and $p = \frac{k}{N}$. ♢

EXAMPLE 1.17.– ***Poisson distribution**: Let $\lambda > 0$ and X be a random variable on $(\Omega, \mathcal{F}, \mathbb{P})$ such that*

$$X(\Omega) = \mathbb{N}$$

and for any $k \in X(\Omega)$,

$$\mathbb{P}(X = k) = \frac{\lambda^k}{k!} e^{-\lambda}.$$

It is then said that X follows a Poisson distribution with parameter λ, and we write $X \sim \mathcal{P}(\lambda)$. ♢

DEFINITION 1.12.– *Let X be a discrete random variable such that $X(\Omega) = \{x_i, i \in I\}$, where $I \subset \mathbb{N}$.*

*– X or the distribution of X is said to be **integrable** (or **summable**) if*

$$\sum_{i \in I} |x_i| \mathbb{P}(X = x_i) < \infty.$$

*– If X is integrable, then the **expectation** of X is the real number defined by*

$$\mathbb{E}[X] = \sum_{i \in I} x_i \mathbb{P}(X = x_i).$$

EXAMPLE 1.18.– *The random variable X defined in Example 1.12 admits an expectation equal to*

$$\begin{aligned}\mathbb{E}[X] &= 10 \times \mathbb{P}(X = 10) + 20 \times \mathbb{P}(X = 20) \\ &= 10 \times \mathbb{P}(\{2, 4, 6\}) + 20 \times \mathbb{P}(\{1, 3, 5\}) \\ &= 10 \times \frac{1}{2} + 20 \times \frac{1}{2} = 15.\end{aligned}$$

The average winnings in the die-rolling game is therefore equal to 15 €. ◇

The following proposition establishes a link between the expectation of a discrete, random variable and measure theory.

PROPOSITION 1.3.– *Let X be a discrete random variable such that $X(\Omega) = \{x_i, i \in I\}$, where $I \subset \mathbb{N}$. It is assumed that $\sum_{i\in I} |x_i|\mathbb{P}(X = x_i) < \infty$. Then,*

$$\mathbb{E}[X] = \int_\Omega X(w)d\mathbb{P}(w).$$

The above proposition also justifies the concept of integrability introduced in Definition 1.12. Further, in this case (i.e. when X is integrable: $\sum_{i\in I} |x_i|\mathbb{P}(X = x_i) < \infty$), we write $X \in L^1(\Omega, \mathcal{F}, \mathbb{P})$.

When X^p is integrable for a certain real number $p \geq 1$ (i.e. when $\sum_{i\in I} |x_i|^p\mathbb{P}(X = x_i) < \infty$), we write

$$X \in L^p(\Omega, \mathcal{F}, \mathbb{P}) \text{ and } \|X\|_p = \left(\mathbb{E}\left[|X|^p\right]\right)^{1/p}.$$

Let us look at some of the properties of expectations.

PROPOSITION 1.4.– *Let X and Y be two integrable, discrete random variables, $a, b \in \mathbb{R}$. Then,*

1) Linearity: $\mathbb{E}[aX + bY] = a\mathbb{E}[X] + b\mathbb{E}[Y]$.

2) Transfer theorem: *if g is a measurable function such that $g(X)$ is integrable, then*

$$\mathbb{E}[g(X)] = \sum_{i\in I} g(x_i)\mathbb{P}(X = x_i).$$

3) Monotonicity: *if $X \leq Y$ almost surely (a.s.), then* $\mathbb{E}[X] \leq \mathbb{E}[Y]$.

4) Cauchy–Schwartz inequality: *If X^2 and Y^2 are integrable, then XY is integrable and*

$$(\mathbb{E}[XY])^2 \leq \mathbb{E}[X^2]\mathbb{E}[Y^2].$$

5) Jensen inequality: *if g is a convex function such that $g(X)$ is integrable, then,*

$$g(\mathbb{E}[X]) \leq \mathbb{E}[g(X)].$$

DEFINITION 1.13.– *Let X be a discrete random variable, such that $X(\Omega) = \{x_i, i \in I\}$, $I \subset \mathbb{N}$ and X^2 is integrable. The* **variance** *of X is the real number:*

$$\mathbb{V}(X) = \mathbb{E}\big[(X - \mathbb{E}[X])^2\big] = \sum_{i \in I}(x_i - \mathbb{E}[X])^2 \mathbb{P}(X = x_i).$$

Variance satisfies the following properties.

PROPOSITION 1.5.– *If a discrete random variable X admits variance, then,*

1) $\mathbb{V}(X) \geq 0$.

2) $\mathbb{V}(X) = \mathbb{E}[X^2] - (\mathbb{E}[X])^2$.

3) For any $(a, b) \in \mathbb{R}^2, \quad \mathbb{V}(aX + b) = a^2\mathbb{V}(X)$.

1.2.3. σ-algebra generated by a random variable

We now define the σ-algebra generated by a random variable. This concept is important for several reasons. For instance, it can make it possible to define the independence of random variables. It is also at the heart of the definition of conditional expectations; see Chapter 2.

PROPOSITION 1.6.– *Let X be a real random variable, defined on $(\Omega, \mathcal{F}, \mathbb{P})$ taking values in $(E, \mathcal{E})$. Then, $\mathcal{F}_X = X^{-1}(\mathcal{E}) = \{X^{-1}(A); A \in \mathcal{E}\}$ is a sub-σ-algebra of $\mathcal{F}$ on Ω. This is called the* **σ-algebra generated by the random variable** *X. It is written as $\sigma(X)$. It is the smallest σ-algebra on Ω that makes X measurable:*

$$\sigma(X) = X^{-1}\left(\mathcal{B}(\mathbb{R})\right) = \{X^{-1}(B); B \in \mathcal{B}(\mathbb{R})\} = \{(X \in B); B \in \mathcal{B}(\mathbb{R})\}.$$

EXAMPLE 1.19.– *Let $\mathcal{F}_0 = \{\emptyset, \Omega\}$ and $X = c \in \mathbb{R}$ be a constant. Then, for any Borel set B in $\mathbb{R}$, $(X \in B)$ has the value $\emptyset$ if $c \notin B$ and Ω if $c \in B$. Thus, the σ-algebra generated by X is $\mathcal{F}_0$. Reciprocally, it can be demonstrated that the only $\mathcal{F}_0$-measurable random variables are the constants. Indeed, let X be a $\mathcal{F}_0$-measurable random variable. Assume that it takes at least two different values, x and y. It may be assumed that $y \geq x$ without loss of generality. Therefore, let $B = [x, \frac{x+y}{2}]$. We have that $(X \in B)$ is non-empty because $x \in B$ but is not Ω since $y \notin B$. Therefore, X is not $\mathcal{F}_0$-measurable.* $\diamondsuit$

PROPOSITION 1.7.– *Let X be a random variable on $(\Omega, \mathcal{F}, \mathbb{P})$ taking values in $(E, \mathcal{E})$ and let $\sigma(X)$ be the σ-algebra generated by X. Thus, a random variable Y is $\sigma(X)$-measurable if and only if there exists a measurable function f such that $Y = f(X)$.*

This technical result will be useful in certain demonstrations further on in the text. In general, if it is known that Y is $\sigma(X)$-measurable, we cannot (and do not need to) make explicit the function f. Reciprocally, if Y can be written as a measurable function of X, it automatically follows that Y is $\sigma(X)$-measurable.

EXAMPLE 1.20.– *A die is rolled 2 times. This experiment is modeled by $\Omega = \{1, 2, 3, 4, 5, 6\}^2$ endowed with the σ-algebra of its subsets and the uniform distribution. Consider the mappings X_1, X_2 and Y from Ω onto $\mathbb{R}$ defined by*

$$X_1(\omega_1, \omega_2) = \omega_1,$$

$$X_2(\omega_1, \omega_2) = \omega_2,$$

$$Y(\omega_1, \omega_2) = \mathbb{1}_{\{2,4,6\}}(\omega_1),$$

thus, X_i is the result of the ith roll and Y is the parity indicator of the first roll. Therefore, $Y = \mathbb{1}_{\{2,4,6\}}(X_1)$; thus, Y is $\sigma(X_1)$-measurable. On the other hand, Y cannot be written as a function of X_2. ◇

The σ-algebra generated by X represents all the events that can be observed by drawing X. It represents the information revealed by X.

DEFINITION 1.14.– *Let $(\Omega, \mathcal{F}, \mathbb{P})$ be a probability space.*

– Let X and Y be two random variables on $(\Omega, \mathcal{F}, \mathbb{P})$ taking values in $(E_1, \mathcal{E}_1)$ and $(E_2, \mathcal{E}_2)$. Then, X and Y are said to be independent if the σ-algebras $\sigma(X)$ and $\sigma(Y)$ are independent.

– Any family $(X_i)_{i \in I}$ of random variables is independent if the σ-algebras $\sigma(X_i)$ are independent.

– Let $\mathcal{G}$ be a sub-σ-algebra of $\mathcal{F}$, and let X be a random variable. Then, X is said to be independent of $\mathcal{G}$ if $\sigma(X)$ is independent of $\mathcal{G}$ or, in other words, $\forall A \in \mathcal{G}$, X and $\mathbb{1}_A$ are independent.

PROPOSITION 1.8.– *If X and Y are two integrable and independent random variables, then their product XY is integrable and $\mathbb{E}[XY] = \mathbb{E}[X]\mathbb{E}[Y]$.*

1.2.4. *Random vectors*

We will now more closely study random variables taking values in $\mathbb{R}^d$, with $d \geq 2$. This concept has already been defined in Definition 1.9. We will now look at the relations between the random vector and its coordinates. When $d = 2$, we then speak of a random couple.

PROPOSITION 1.9.– *Let X be a real random vector on the probability space $(\Omega, \mathcal{F}, \mathbb{P})$, taking values in $\mathbb{R}^d$. Then,*

$$X(w) = \begin{pmatrix} X_1(w) \\ \vdots \\ X_n(w) \end{pmatrix}$$

is such that for any $i \in \{1, ..., d\}$, X_i is a real random variable.

DEFINITION 1.15.– *A random vector is said to be discrete if each of its components, X_i, is a discrete random variable.*

DEFINITION 1.16.– *Let $X = \begin{pmatrix} X_1 \\ X_2 \end{pmatrix}$ be a discrete random couple such that*

$$X_1(\Omega) = \{x_{1j}, j \in I_1\} \text{ et } X_2(\Omega) = \{x_{2k}, k \in I_2\}.$$

*The **conjoint distribution** (or **joint distribution** or, simply, the distribution) of X is given by the family*

$$\{\mathbb{P}(X_1 = x_{1j}, X_2 = x_{2k});\ (j,k) \in I_1 \times I_2\}.$$

*The **marginal distributions** of X are the distributions of X_1 and X_2. These distributions may be derived from the conjoint distribution of X through:*

$$\forall j \in I_1,\ \ \mathbb{P}(X_1 = x_{1j}) = \sum_{k \in I_2} \mathbb{P}(X_1 = x_{1j}, X_2 = x_{2k})$$

and

$$\forall k \in I_2,\ \ \mathbb{P}(X_2 = x_{2k}) = \sum_{j \in I_1} \mathbb{P}(X_1 = x_{1j}, X_2 = x_{2k}).$$

The concept of joint distributions and marginal distributions can naturally be extended to vectors with dimension larger than 2.

EXAMPLE 1.21.– *A coin is tossed 3 times, and the result is noted. The universe of possible outcomes is $\Omega = \{T, H\}^3$. Let X denote the total number of tails obtained and Y denote the number of tails obtained at the first toss. Then,*

$$X(\Omega) = \{0, 1, 2, 3\} \text{ and } Y(\Omega) = \{0, 1\}.$$

The couple (X, Y) is, therefore, a random vector (referred to here as a "random couple"), with joint distribution defined by

$$\mathbb{P}((X,Y)=(i,j)) = \begin{cases} \dfrac{1}{2^3} \; if \; (i,j) \in \{(0,0),(1,1),(2,0),(3,1)\} \\ 2 \times \dfrac{1}{2^3} \; if \; (i,j) \in \{(1,0),(2,1)\} \\ 0 \; if \; (i,j) \in \{(3,0),(0,1)\}, \end{cases}$$

for any $(i,j) \in X(\Omega) \times Y(\Omega)$, which makes it possible to derive the distributions of X and Y (called the marginal distributions of the couple (X,Y)):

Distribution of X:

$$\mathbb{P}(X=0) = \mathbb{P}(X=0, Y=0) + \mathbb{P}(X=0, Y=1) = \frac{1}{2^3}$$

$$\mathbb{P}(X=1) = \mathbb{P}(X=1, Y=0) + \mathbb{P}(X=1, Y=1) = \frac{3}{2^3}$$

$$\mathbb{P}(X=2) = \mathbb{P}(X=2, Y=0) + \mathbb{P}(X=2, Y=1) = \frac{3}{2^3}$$

$$\mathbb{P}(X=3) = \mathbb{P}(X=3, Y=0) + \mathbb{P}(X=3, Y=1) = \frac{1}{2^3}.$$

Distribution of Y:

$$\mathbb{P}(Y=0) = \sum_{i=0}^{3} \mathbb{P}(X=i, Y=0) = 4 \times \frac{1}{2^3} = \frac{1}{2}$$

$$\mathbb{P}(Y=1) = \sum_{i=0}^{3} \mathbb{P}(X=i, Y=1) = 4 \times \frac{1}{2^3} = \frac{1}{2}.$$

◇

1.2.5. *Convergence of sequences of random variables*

To conclude this section on random variables, we will review some classic results of convergence for sequences of random variables. Throughout the rest of this book, the abbreviation *r.v.* signifies *random variable*.

DEFINITION 1.17.– *Let $(X_n)_{n \geq 1}$ and X be r.v.s defined on $(\Omega, \mathcal{F}, \mathbb{P})$.*

1) It is assumed that there exists $p > 0$ such that, for any $n \geq 0$, $\mathbb{E}[|X_n|^p] < \infty$, and $\mathbb{E}[|X|^p] < \infty$. It is said that the sequence of random variables $(X_n)_{n \geq 1}$ ***converges on the average of the order p*** *or* ***converges in L^p*** *towards X, if*

$$\lim_{n \to \infty} \mathbb{E}[|X_n - X|^p] = 0.$$

We then write $X_n \xrightarrow{L^p} X$. In the specific case where $p = 2$, we say there is a ***convergence in quadratic mean.***

2) The sequence of r.v. $(X_n)_{n\geq 1}$ is called ***almost surely (a.s.) convergent*** *towards X, if*

$$\lim_{n\to\infty} \mathbb{P}\left(w \in \Omega;\ \lim_{n\to\infty} X_n(w) = X(w)\right) = 1.$$

We then write $X_n \xrightarrow{a.s.} X$.

THEOREM 1.1 (Monotone convergence theorem).– *Let $(X_n)_{n\geq 1}$ be a sequence of positive and non-decreasing random variables and let X be an integrable random variable, all of these defined on the same probability space $(\Omega, \mathcal{F}, \mathbb{P})$. If (X_n) converges almost surely to X, then*

$$\lim_{n\to+\infty} \mathbb{E}[X_n] = \mathbb{E}[X].$$

THEOREM 1.2 (Dominated convergence theorem).– *Let $(X_n)_{n\geq 1}$ be a sequence of random variables and let X be another random variable, all defined on the same probability space $(\Omega, \mathcal{F}, \mathbb{P})$. If the sequence (X_n) converges to X a.s., and for any $n \geq 1$, $|X_n| \leq Z$, where Z is an integrable random variable, then $X_n \xrightarrow{L^1} X$ and, in particular,*

$$\lim_{n\to+\infty} \mathbb{E}[X_n] = \mathbb{E}[X].$$

THEOREM 1.3 (Strong law of large numbers).– *Let $(X_n)_{n\geq 1}$ be a sequence of integrable, independent random variables from the same distribution. Then,*

$$\frac{X_1 + X_2 + \cdots + X_n}{n} \xrightarrow[n\to\infty]{a.s.} \mathbb{E}[X_1].$$

1.3. Stochastic processes

The main objective of this book is to study certain families of stochastic (or random) processes in discrete time. There are two ways of seeing such objects:

– as a sequence $(X_n)_{n\in\mathbb{N}}$ of real random variables;

– as a single random variable X taking values in the set of real sequences.

The index n represents time. Since $n \in \mathbb{N}$, we speak of processes in discrete time. In the rest of this book, unless indicated otherwise, we will only consider processes taking discrete real values. The notation E thus denotes a finite or countable subset of $\mathbb{R}$ and $\mathcal{E} = \mathcal{P}(E)$, the set of subsets of E.

DEFINITION 1.18.– *A* ***stochastic process*** *is a sequence $X = (X_n)_{n\in\mathbb{N}}$ of random variables taking values in $(E, \mathcal{E})$. The process X is then a random variable taking values in $(E^{\mathbb{N}}, \mathcal{E}^{\otimes\mathbb{N}})$.*

EXAMPLE 1.22.– *A coin is tossed an infinite number of times. This experiment is modeled by* $\Omega = \{T, H\}^{\mathbb{N}^*}$. *For* $n \in \mathbb{N}^*$, *consider the mappings* X_n *to* Ω *in* $\mathbb{R}$ *defined by*

$$X_n(\omega_1, \omega_2, \ldots, \omega_n, \ldots) = \mathbb{1}_{\{T\}}(\omega_n),$$

the number of tails at the n*th toss. Therefore,* X_n, $n \in \mathbb{N}^*$ *are discrete, real random variables and the sequence* $X = (X_n)_{n\in\mathbb{N}}$ *is a stochastic process.* ◇

DEFINITION 1.19.– *Let* $X = (X_n)_{n\in\mathbb{N}}$ *be a stochastic process. For all* $n \in \mathbb{N}$, *the distribution of the vector* $(X_0, X_1, \ldots, X_n)$ *is denoted by* μ_n. *The probability distributions* $(\mu_n)_{n\in\mathbb{N}}$ *are called* ***finite-dimensional distributions*** *or* ***finite-dimensional marginal distributions*** *of the process* $X = (X_n)_{n\in\mathbb{N}}$.

PROPOSITION 1.10.– *Let* $X = (X_n)_{n\in\mathbb{N}}$ *be a stochastic process and let* $(\mu_n)_{n\in\mathbb{N}}$ *be its finite-dimensional distributions. Then, for all* $n \in \mathbb{N}^*$ *and* $(A_0, \ldots, A_{n-1}) \in \mathcal{E}^n$, *we have*

$$\mu_{n-1}(A_0 \times \ldots \times A_{n-1}) = \mu_n(A_0 \times \ldots \times A_{n-1} \times E).$$

In other words, the restriction of the marginal distribution of the vector $(X_0, \ldots, X_n)$ *to its first* n *coordinates is exactly the distribution of the vector* $(X_0, \ldots, X_{n-1})$.

PROOF.– This proof directly follows from the definition of the objects. We have

$$\begin{aligned}\mu_{n-1}(A_0 \times \ldots \times A_{n-1}) &= \mathbb{P}(X_0 \in A_0, \ldots, X_{n-1} \in A_{n-1}) \\ &= \mathbb{P}(X_0 \in A_0, \ldots, X_{n-1} \in A_{n-1}, X_n \in E) \\ &= \mu_n(A_0 \times \ldots \times A_{n-1} \times E),\end{aligned}$$

and hence, the desired equality. □

Indeed, this property completely characterizes the distribution of the process X according to the following theorem.

THEOREM 1.4 (Kolmogorov).– *The* **canonical space** $(\Omega, \mathcal{F})$ *is defined in the following manner. Let* $\Omega = E^{\mathbb{N}}$. *The coordinate mappings* $(X_n)_{n\in\mathbb{N}}$ *are defined by* $X_n(\omega) = \omega_n$ *for any* $\omega = (\omega_n)_{n\in\mathbb{N}} \in \Omega$ *and we write* $\mathcal{F} = \sigma(X_n, n \in \mathbb{N})$. *Let* $(\mu_n)_{n\in\mathbb{N}}$ *be a family of probability distributions such that*

1) for any $n \in \mathbb{N}$, μ_n *is defined on* $(E^{n+1}, \mathcal{E}^{\otimes(n+1)})$,

2) for any $n \in \mathbb{N}^*$ *and* $(A_0, \ldots, A_{n-1}) \in \mathcal{E}^n$, *we have* $\mu_{n-1}(A_0 \times \ldots \times A_{n-1}) = \mu_n(A_0 \times \ldots \times A_{n-1} \times E)$.

Therefore, there exists a unique probability distribution μ over the canonical space $(\Omega, \mathcal{F})$ such that the process $X = (X_n)_{n \in \mathbb{N}}$ for the coordinate mapping has the distribution μ and for the finite-dimensional distributions has the sequence $(\mu_n)_{n \in \mathbb{N}}$.

This result is very important for the theory of processes as it signifies that it is sufficient to specify (all) the finite-dimensional distributions and for them to be compatible with each other, to uniquely define a process distribution over the space of infinite random sequences. In practice, this makes it possible to justify the construction of processes (existence property) as well as showing that two processes have the same distribution (unicity property).

To study the random variables taking values in the set of sequences, we need new definitions for σ-algebras and measurability.

DEFINITION 1.20.– *In a probability space $(\Omega, \mathcal{F}, \mathbb{P})$, a **filtration** is a sequence $(\mathcal{F}_n)_{n \in \mathbb{N}}$ of sub-σ-algebras of $\mathcal{F}$ such that, for any $n \in \mathbb{N}$, $\mathcal{F}_n \subset \mathcal{F}_{n+1}$. This is, thus, a non-decreasing sequence (for inclusion) of sub-σ-algebras of $\mathcal{F}$.*

*When $(\mathcal{F}_n)_{n \in \mathbb{N}}$ is a filtration defined on the probability space $(\Omega, \mathcal{F}, \mathbb{P})$, the quadruplet $(\Omega, \mathcal{F}, \mathbb{P}, (\mathcal{F}_n)_{n \in \mathbb{N}})$ is said to be a **filtered probability space**.*

EXAMPLE 1.23.– *Let $(X_n)_{n \in \mathbb{N}}$ be a sequence of random variables and we consider, for any $n \in \mathbb{N}$, $\mathcal{F}_n = \sigma(X_0, X_1, ..., X_n)$, the σ-algebra generated by $\{X_0, ..., X_n\}$. The sequence $(\mathcal{F}_n)_{n \in \mathbb{N}}$ is, therefore, a filtration, called a **natural filtration** of $(X_n)_{n \in \mathbb{N}}$ or filtration generated by $(X_n)_{n \in \mathbb{N}}$. This filtration represents the information revealed over time, by the observation of the drawings of the sequence $X = (X_n)_{n \in \mathbb{N}}$.* ◇

DEFINITION 1.21.– *Let $(\Omega, \mathcal{F}, \mathbb{P}, (\mathcal{F}_n)_{n \in \mathbb{N}})$ be a filtered probability space, and let $X = (X_n)_{n \in \mathbb{N}}$ be a stochastic process.*

*– X is said to be **adapted** to the filtration $(\mathcal{F}_n)_{n \in \mathbb{N}}$ (or again $(\mathcal{F}_n)_{n \in \mathbb{N}}$–adapted), if X_n is $\mathcal{F}_n$-measurable for any $n \in \mathbb{N}$;*

*– X is said to be **predictable** with respect to the filtration $(\mathcal{F}_n)_{n \in \mathbb{N}}$ (or again $(\mathcal{F}_n)_{n \in \mathbb{N}}$–predictable), if X_n is $\mathcal{F}_{n-1}$-measurable for any $n \in \mathbb{N}^*$.*

EXAMPLE 1.24.– *A process is always adapted with respect to its natural filtration.* ◇

As its name indicates, for a predictable process, we know its value X_n from the instant $n - 1$.

1.4. Exercises

EXERCISE 1.1.– Let $\Omega = \{a, b, c\}$.

1) Completely describe all the σ-algebras of Ω.

2) State which are the sub-σ-algebras of which.

EXERCISE 1.2.– Let $\Omega = \{a, b, c, d\}$. Among the following sets, which are σ-algebras?

1) $\mathcal{A}_1 = \{\emptyset, \Omega, \{b\}, \{a, c, d\}\}$;

2) $\mathcal{A}_2 = \{\emptyset, \Omega, \{b, c, d\}, \{c, d\}\}$;

3) $\mathcal{A}_3 = \{\{b\}\}$;

4) $\mathcal{A}_4 = \{\emptyset, \Omega, \{a, c\}, \{b, d\}\}$.

For those which are not σ-algebras, completely describe the σ-algebras they generate.

EXERCISE 1.3.– Let X be a random variable on $(\Omega, \mathcal{F})$ and let $\mathcal{G}$ be a sub-σ-algebra of $\mathcal{F}$. Show that X is $\mathcal{G}$-measurable if and only if $\sigma(X) \subset \mathcal{G}$.

EXERCISE 1.4.– Let $A \in \mathcal{F}$ and let $\mathcal{G}$ be a sub-σ-algebra of $\mathcal{F}$. Show that $\mathbb{1}_A$ is $\mathcal{G}$-measurable if and only if $A \in \mathcal{G}$.

EXERCISE 1.5.– Let $\Omega = \{P, F\} \times \{P, F\}$ and $\mathcal{F} = \mathcal{P}(\Omega)$, corresponding to two successive coin tosses. Let

– X_1 be the random variable *number of T on the first toss*;

– X_2 be *the number of T on the second toss*;

– Y be *the number of T obtained on the two tosses*;

– and $Z = 1$ if the two tosses yielded an identical result; otherwise, it is 0.

1) Describe $\mathcal{F}_1 = \sigma(X_1)$ and $\mathcal{F}_2 = \sigma(X_2)$. Is X_1 $\mathcal{F}_2$-measurable?

2) Describe $\mathcal{G} = \sigma(Y)$. Is Y $\mathcal{F}_1$-measurable? Is X_1 $\mathcal{G}$-measurable?

3) Describe $\mathcal{H} = \sigma(Z)$. Is Z $\mathcal{F}_1$-measurable, $\mathcal{G}$-measurable? Is X_1 $\mathcal{H}$-measurable?

4) Give the inclusions between $\mathcal{F}$, $\mathcal{F}_1$, $\mathcal{F}_2$, $\mathcal{G}$ and $\mathcal{H}$.

EXERCISE 1.6.– Let $(X_n)_{n\geq 1}$ be a sequence of independent random variables with the same Rademacher distribution with parameter $1/2$:

$$\mathbb{P}(X_1 = 1) = \mathbb{P}(X_1 = -1) = 1/2.$$

Let $S_0 = 0$ and $S_n = \sum_{k=1}^{n} X_k$ for $n \geq 1$. The process $S = (S_n)_{n\in\mathbb{N}}$ is called the simple symmetric random walk on $\mathbb{Z}$. It will be studied in detail in Chapter 3. We write $X_0 = 0$. Show that the filtration generated by the sequence $(X_n)_{n\in\mathbb{N}}$ is the same as that generated by the sequence $(S_n)_{n\in\mathbb{N}}$.

EXERCISE 1.7.– Consider the following game of chance. A player begins by choosing a number between 6 and 8 (inclusive), which we call the principal. The player then rolls 2 uncut, six-sided, non-rigged dice and sums the result. The wins are as follows:

– If the sum is 2 or 3, the player loses 1 DT (Tunisian dinar).

– If the sum is 11, the player wins 1 DT if the principal is 7; otherwise, they lose 1 DT.

– If the sum is 12, the player wins 1 DT if the principal is 6 or 8; otherwise, they lose 1 DT.

– Finally, in all other cases, nothing happens (no win, no loss).

1) Determine Ω, the universe of all outcomes of the experiment.

2) S is the random variable giving the sum of the two dice. Determine the distribution of S.

3) X_7 is the random variable giving the player's winnings (negative in the case of a loss) when the principal is 7. Determine the distribution of X_7.

4) Calculate $\mathbb{E}[X_7]$ and $\mathbb{V}(X_7)$.

5) We consider X_6 to be the winnings of the player when the principal is 6. Determine the distribution of X_6 and then calculate $\mathbb{E}[X_6]$.

EXERCISE 1.8.– Let $p \in]0, 1[$. We have one coin that leads to tails with the probability p. We toss this coin until we obtain tails for the second time. Let X be the number of heads obtained during this experiment.

1) Determine the distribution of X.

2) Justify the existence of the expectation of X.

3) Calculate $\mathbb{E}[X]$.

2

Conditional Expectation

This chapter is dedicated to the concept of conditional expectation. This is a very important concept that generalizes the concept of conditional probability, when conditioning occurs not only on a single event, but on an entire family of events (a σ-algebra). It is also the basic concept to define martingales, as we will see in Chapter 4. The readers who may wish to explore the concepts in this chapter in greater detail can consult [BAR 07, BIL 12, CHU 01, DAC 83, FOA 03, KAL 02, OUV 08, OUV 09].

Let us begin in section 2.1 by reviewing the definition of conditional probability with respect to an event with strictly positive probability. Following this, section 2.2 gives the general definition for conditional expectations and its chief properties. Next, section 2.3 gives a geometric interpretation as an orthogonal projection in the particular cases of square integrable random variables. Section 2.4 explores the relations between conditional expectation and independence. Finally, section 2.5 offers exercises in working with all these concepts.

Throughout this chapter, we will work on a probability space $(\Omega, \mathcal{F}, \mathbb{P})$.

2.1. Conditional probability with respect to an event

Conditional probability accounts for the information brought in by the occurrence of one event on the probability of the occurrence of another event.

DEFINITION 2.1.– *Let $B \in \mathcal{F}$ such that $\mathbb{P}(B) > 0$. The **conditional probability** of the event A given the event B is the number*

$$\mathbb{P}(A|B) = \frac{\mathbb{P}(A \cap B)}{\mathbb{P}(B)}.$$

EXAMPLE 2.1.– *There are two roads to go from city A to city B and two roads from city B to city C. Each of these roads is blocked by snow, with a probability p, independently of the others. We wish to find the probability that there is a practicable road from A to B given that it is not possible to travel between A and C. We introduce the following events*

AB: at least one road is open between A and B,
BC: at least one road is open between B and C.

We have $\mathbb{P}(AB) = \mathbb{P}(BC) = 1 - p^2$ since the two possible roads are blocked independently of one another. Further, the events AB and BC are independent. We wish to find $\mathbb{P}(AB|(AB \cap BC)^c)$. We have, using independence,

$$\begin{aligned}
\mathbb{P}(AB|(AB \cap BC)^c) &= \frac{\mathbb{P}(AB \cap (AB \cap BC)^c)}{\mathbb{P}((AB \cap BC)^c)} \\
&= \frac{\mathbb{P}(AB \cap (AB^c \cup BC^c))}{1 - \mathbb{P}(AB \cap BC)} \\
&= \frac{\mathbb{P}((AB \cap AB^c) \cup (AB \cap BC^c))}{1 - \mathbb{P}(AB)\mathbb{P}(BC)} \\
&= \frac{\mathbb{P}(AB \cap BC^c)}{1 - \mathbb{P}(AB)\mathbb{P}(BC)} \\
&= \frac{\mathbb{P}(AB)\mathbb{P}(BC^c)}{1 - \mathbb{P}(AB)\mathbb{P}(BC)} \\
&= \frac{(1-p^2)p^2}{1-(1-p^2)^2}.
\end{aligned}$$

The probability that there is a practicable route from A to B given that we cannot travel between A and C is thus $\frac{(1-p^2)p^2}{1-(1-p^2)^2}$. ◇

PROPOSITION 2.1.– *Let $B \in \mathcal{F}$ such that $\mathbb{P}(B) > 0$. The mapping $\mathbb{P}(\cdot|B)$ from $\mathcal{F}$ to $\mathbb{R}^+$ defined by*

$$A \longmapsto \mathbb{P}(A|B)$$

is a probability over $(\Omega, \mathcal{F})$. It is called the **conditional distribution** *given B.*

PROOF.– It is enough to verify that it is a σ-additive mapping that satisfies $\mathbb{P}(\Omega|B) = 1$. On the one hand, we have $\mathbb{P}(\Omega|B) = \mathbb{P}(B)/\mathbb{P}(B) = 1$ and if $(A_n)_{n \in \mathbb{N}}$ is a sequence of pairwise disjoint elements of $\mathcal{F}$, then $(A_n \cap B)_{n\in\mathbb{N}}$ is also a sequence of pairwise disjoint elements; hence,

$$\mathbb{P}(\bigcup_{n\in\mathbb{N}} A_n|B) = \frac{\mathbb{P}\left(\bigcup_{n\in\mathbb{N}} A_n \cap B\right)}{\mathbb{P}(B)} = \frac{\sum_{n\in\mathbb{N}} \mathbb{P}(A_n \cap B)}{\mathbb{P}(B)} = \sum_{n\in\mathbb{N}} \mathbb{P}(A_n|B),$$

and hence the stated result. □

Let X be a random variable on $(\Omega, \mathcal{F}, \mathbb{P})$. Let $B \in \mathcal{F}$ be an event such that $\mathbb{P}(B) > 0$. We now have a second probability distribution over Ω, the conditional distribution given B. Thus, for the random variable, X, it is possible to define its *conditional distribution given* B in the following manner.

DEFINITION 2.2.– *Let X be a random variable on $(\Omega, \mathcal{F}, \mathbb{P})$ taking values in $(E, \mathcal{E})$. The* ***conditional distribution of*** X ***given*** B*, denoted by $\mathbb{P}_{X|B}$, is the image measure of $\mathbb{P}(\cdot|B)$ under X:*

$$\forall A \in \mathcal{E}, \quad \mathbb{P}_{X|B}(A) = \mathbb{P}(X \in A|B).$$

EXAMPLE 2.2.– *Consider the experiment of rolling a die; the random variable X is equal to the result of the rolling of the die, and consider the event B:* "the result is even". *The distribution of X is the uniform distribution over $\{1, 2, 3, 4, 5, 6\}$. The distribution of X given B is*

$$\mathbb{P}(X = 1|B) = \frac{\mathbb{P}((X = 1) \cap B)}{\mathbb{P}(B)} = 0$$

$$\mathbb{P}(X = 2|B) = \frac{\mathbb{P}((X = 2) \cap B)}{\mathbb{P}(B)} = \frac{\mathbb{P}(X = 2)}{\mathbb{P}(B)} = \frac{2}{6} = \frac{1}{3}$$

$$\mathbb{P}(X = 3|B) = \frac{\mathbb{P}((X = 3) \cap B)}{\mathbb{P}(B)} = 0$$

$$\mathbb{P}(X = 4|B) = \frac{\mathbb{P}((X = 4) \cap B)}{\mathbb{P}(B)} = \frac{\mathbb{P}(X = 4)}{\mathbb{P}(B)} = \frac{2}{6} = \frac{1}{3}$$

$$\mathbb{P}(X = 5|B) = \frac{\mathbb{P}((X = 5) \cap B)}{\mathbb{P}(B)} = 0$$

$$\mathbb{P}(X = 6|B) = \frac{\mathbb{P}((X = 6) \cap B)}{\mathbb{P}(B)} = \frac{\mathbb{P}(X = 6)}{\mathbb{P}(B)} = \frac{2}{6} = \frac{1}{3}$$

Therefore, the distribution of X given B is the uniform distribution over $\{2, 4, 6\}$. ◇

We now characterize the expectation for the conditional distribution of X given B, when this expectation exists.

PROPOSITION 2.2.– *If $X \in L^1(\Omega, \mathcal{F}, \mathbb{P})$, then $X \in L^1(\Omega, \mathcal{F}, \mathbb{P}(\cdot|B))$ and if $\mathbb{E}[X|B]$ denotes the expectation of X for the conditional distribution $\mathbb{P}(\cdot|B)$, we then have*

$$\mathbb{E}[X|B] = \frac{1}{\mathbb{P}(B)} \int_B X(\omega) d\mathbb{P}(\omega) = \frac{1}{\mathbb{P}(B)} \int_B x d\mathbb{P}_X(x) = \frac{1}{\mathbb{P}(B)} \mathbb{E}[X \mathbb{1}_B].$$

PROOF.– Given the definition of the conditional distribution $\mathbb{P}(\cdot|B) = \mathbb{P}(\cdot \cap B)/\mathbb{P}(B)$, we have

$$\int_\Omega |X(\omega)| d\mathbb{P}(\omega|B) = \frac{1}{\mathbb{P}(B)} \int_B |X(\omega)| d\mathbb{P}(\omega) \leq \frac{1}{\mathbb{P}(B)} \int_\Omega |X(\omega)| d\mathbb{P}(\omega) < \infty,$$

if $X \in L^1(\Omega, \mathcal{F}, \mathbb{P})$, then X is indeed integrable for the distribution $\mathbb{P}(\cdot|B)$. In this case, we have

$$\mathbb{E}[X|B] = \frac{1}{\mathbb{P}(B)} \int_B X(\omega) d\mathbb{P}(\omega) = \frac{1}{\mathbb{P}(B)} \int_B x d\mathbb{P}_X(x) = \frac{1}{\mathbb{P}(B)} \mathbb{E}[X \mathbb{1}_B],$$

and hence, the result. □

EXAMPLE 2.3.– *We return to the die-rolling example given above. We have $\mathbb{E}[X] = 7/2$ and $\mathbb{E}[X|B] = (2+4+6)/3 = 4$. We also have $\mathbb{E}[X\mathbb{1}_B] = (2+4+6)/6 = 2$ and $\mathbb{P}(B) = 1/2$, and we arrive at $\mathbb{E}[X|B] = 4$.* ◇

2.2. Conditional expectation

In the following sections, we wish to expand the concept of conditioning not with respect to a single event, but with respect to an entire family of events. We also wish to go beyond the hypothesis that $\mathbb{P}(B) > 0$. In order to perform this, we first generalize the concept of conditional expectation.

2.2.1. *Definitions*

The theorem given below broadens the definition of conditional expectation when we condition with respect to a σ-algebra.

THEOREM 2.1.– *Let X be a random variable in $L^1(\Omega, \mathcal{F}, \mathbb{P})$ and let $\mathcal{G}$ be a sub-σ-algebra of $\mathcal{F}$. Then, there exists a unique random variable $Y \in L^1(\Omega, \mathcal{F}, \mathbb{P})$ such that*

1) Y is $\mathcal{G}$-measurable;

2) for any $A \in \mathcal{G}$, we have $\mathbb{E}[X\mathbb{1}_A] = \mathbb{E}[Y\mathbb{1}_A]$.

The unicity is intended almost surely, that is, if two random variables Y and Z satisfy the above properties, then $\mathbb{P}(Y = Z) = 1$. Informally, Y is a random, $\mathcal{G}$-measurable variable, which *resembles* X in the sense that it coincides with X on $\mathcal{G}$. In a sense, which will be explained in detail in section 2.3, it is the best approximation of X that is $\mathcal{G}$-measurable. In other words, Y is the best prediction of X when the quantity of information revealed by the σ-algebra $\mathcal{G}$ is available.

PROOF.– Existence and unicity are demonstrated separately.

Existence: first case: assume that $X \geq 0$. We apply the Radon–Nykodym theorem[1] with $\mu = \mathbb{P}$ which is a measure on the probability space $(\Omega, \mathcal{G})$ since $\mathcal{G}$ is a sub-σ-algebra of $\mathcal{F}$. As regards ν, consider the measure defined over $(\Omega, \mathcal{G})$ by:

$$\forall B \in \mathcal{G},\ \nu(B) = \mathbb{E}[X\mathbb{1}_B] = \int_B X d\mathbb{P}.$$

We then have $\nu \ll \mu$. Indeed, we have

$$\mathbb{P}(A) = 0 \Longrightarrow \mathbb{1}_A = 0\ \mathbb{P} - a.s. \Longrightarrow X\mathbb{1}_A = 0\ \mathbb{P} - a.s. \Longrightarrow \nu(A) = 0.$$

Thus, using the Radon–Nykodym theorem, there exists a function Y that is $\mathcal{G}$-measurable (i.e. a random $\mathcal{G}$-measurable variable Y), such that for any $A \in \mathcal{G}$, we have $\nu(A) = \displaystyle\int_A Y d\mathbb{P}$, which translates into the desired equality $\mathbb{E}[X\mathbb{1}_A] = \mathbb{E}[Y\mathbb{1}_A]$.

Second case: if the random variable X has any sign, it is decomposed as the difference of its positive part and its negative part: $X = X^+ - X^-$, where

$$X^+ = \max(X, 0)\ and\ X^- = \max(-X, 0).$$

As $X^+ \geq 0$ and $X^- \geq 0$, the first case can be applied to them: there exist two random, $\mathcal{G}$-measurable variables, Y^+ and Y^-, such that for any $A \in \mathcal{G}$, we have $\mathbb{E}[X^+\mathbb{1}_A] = \mathbb{E}[Y^+\mathbb{1}_A]$ and $\mathbb{E}[X^-\mathbb{1}_A] = \mathbb{E}[Y^-\mathbb{1}_A]$. The desired result is obtained, therefore, by writing $Y = Y^+ - Y^-$.

---

1 **Radon–Nykodym theorem:** let μ and ν be two $\sigma-$ finite measures on a probability space $(\Omega, \mathcal{F})$, such that $\nu \ll \mu$ (ν is absolutely continuous with respect to μ, that is, for any $A \in \mathcal{F}$, $\mu(A) = 0 \Longrightarrow \nu(A) = 0$). Then, there exists f, an $\mathcal{F}$-measurable function, such that for any $A \in \mathcal{F}$, we have

$$\nu(A) = \int_A f d\mu.$$

The function f is then called the **Radon–Nykodym derivative** of ν with respect to μ and is denoted by $f = \dfrac{d\nu}{d\mu}$.

Unicity: let Y_1 and Y_2 satisfy statements 1 and 2 of Theorem 2.1. We have, in particular

$$\forall A \in \mathcal{G},\ \mathbb{E}[Y_1 \mathbb{1}_A] = \mathbb{E}[Y_2 \mathbb{1}_A].$$

For a fixed $\epsilon > 0$, we write $A_\epsilon = \{Y_1 - Y_2 \geq \epsilon\}$. As $Y_1 - Y_2$ is $\mathcal{G}-$ measurable, we have that $A_\epsilon \in \mathcal{G}$. It follows then that

$$0 = \mathbb{E}[(Y_1 - Y_2)\mathbb{1}_{A_\epsilon}] \geq \epsilon\mathbb{E}[\mathbb{1}_{A_\epsilon}] = \epsilon\mathbb{P}(A_\epsilon),$$

which implies that

$$\mathbb{P}(A_\epsilon) = 0.$$

Now, let $A^+ = \{Y_1 - Y_2 > 0\} = \bigcup_{n\geq 1} A_{1/n}$. Because $(A_{1/n})_{n\geq 1}$ is an increasing sequence, according to Proposition 1.2, we have

$$\mathbb{P}(A^+) = \mathbb{P}\Big(\bigcup_{n\geq 1} A_{1/n}\Big) = \lim_{n\to\infty} \mathbb{P}(A_{1/n}) = 0.$$

By switching the roles of Y_1 and Y_2, if we write $A^- = \{Y_1 - Y_2 < 0\}$, then we also have $\mathbb{P}(A^-) = 0$. Therefore, since $\{Y_1 \neq Y_2\} = A^+ \cup A^-$, we have $\mathbb{P}(\{Y_1 \neq Y_2\}) = 0$ and consequently, $Y_1 = Y_2$ a.s. □

DEFINITION 2.3.– *The random variable Y defined in Theorem 2.1 is called the* ***conditional expectation*** *of X given $\mathcal{G}$ and is denoted by $\mathbb{E}[X|\mathcal{G}]$.*

REMARK 2.1.– *This definition of conditional expectation is based on an existence theorem and is not constructive. It does not make it possible to directly explicitize conditional expectation. However, as will be seen in the following section, from this abstract definition, it is possible to obtain some properties of conditional expectation which will be enough to carry out the calculations in this book. However, in some special cases where the σ-algebra is generated by a discrete family, it is still possible to carry out an explicit calculation.* ★

EXAMPLE 2.4.– *Let $(B_i)_i$ be a partition of Ω such that for any i, $\mathbb{P}(B_i) > 0$ and let X be an integrable, random variable. Let $\mathcal{G} = \sigma(B_1, B_2, \ldots)$ be the σ-algebra generated by B_i. Then, $\mathbb{E}[X|\mathcal{F}]$ is the random variable defined by*

$$\mathbb{E}[X|\mathcal{G}](w) = \sum_i \frac{\mathbb{E}[X\mathbb{1}_{B_i}]}{\mathbb{P}(B_i)} \mathbb{1}_{B_i}(w) = \mathbb{1}_{B_i}(w)\frac{1}{\mathbb{P}(B_i)} \int_{B_i} X d\mathbb{P}.$$

Indeed, let us write $Y = \sum_i \frac{\mathbb{E}[X\mathbb{1}_{B_i}]}{\mathbb{P}(B_i)} \mathbb{1}_{B_i}$. Then, Y is a linear combination of the $\mathbb{1}_{B_i}$ which are all $\mathcal{G}$-measurable. Thus, Y is indeed $\mathcal{G}$-measurable. Since $\mathcal{G}$ is the

σ-algebra generated by $B_1, B_2, \ldots$, in order to satisfy point 2 in Theorem 2.1, it is sufficient to verify it for $A = B_k$, for any k. Further, we have

$$\begin{aligned}\mathbb{E}[Y\mathbb{1}_{B_k}] &= \mathbb{E}\Big[\sum_i \frac{\mathbb{E}[X\mathbb{1}_{B_i}]}{\mathbb{P}(B_i)}\mathbb{1}_{B_k}\mathbb{1}_{B_i}\Big] \\ &= \mathbb{E}\Big[\frac{\mathbb{E}[X\mathbb{1}_{B_k}]}{\mathbb{P}(B_k)}\mathbb{1}_{B_k}\Big] \\ &= \mathbb{E}[X\mathbb{1}_{B_k}],\end{aligned}$$

as the B_i are pairwise disjoint, hence the result. It must also be noted that this conditional expectation can also be re-written as

$$\mathbb{E}[X|\mathcal{G}] = \sum_i \frac{\mathbb{E}[X\mathbb{1}_{B_i}]}{\mathbb{P}(B_i)}\mathbb{1}_{B_i} = \sum_i \mathbb{E}[X|B_i]\mathbb{1}_{B_i},$$

by using Proposition 2.2. This result remains true for a complete system of events, that is, a family $(B_i)_i$ such that $\cup_i B_i = \Omega$ and $\mathbb{P}(B_i \cap B_j) = 0$ for any $i \neq j$. ◇

DEFINITION 2.4.– *Let X and Y be two r.v.s with $X \in L^1(\Omega, \mathcal{F}, \mathbb{P})$. The expectation of X given Y is defined by:*

$$\mathbb{E}[X|Y] = \mathbb{E}[X|\sigma(Y)],$$

where $\sigma(Y)$ is the σ-algebra generated by the random variable Y.

EXAMPLE 2.5.– *When Y is a discrete random variable, then $\mathbb{E}[X|Y]$ may be calculated according to the method introduced in the preceding example. To be more precise, it is assumed that $Y(\Omega) = \{y_i; i \geq 1\}$ and it is assumed that for any $i \geq 1$, $\mathbb{P}(Y = y_i) > 0$. Then, $\mathbb{E}[X|Y] = h(Y)$ with $h : Y(\Omega) \longrightarrow \mathbb{R}$ defined by:*

$$y \in Y(\Omega) \longmapsto h(y) = \frac{\mathbb{E}[X\mathbb{1}_{Y=y}]}{\mathbb{P}(Y = y)}.$$

In other words, we have

$$\mathbb{E}[X|Y] = \sum_{n=1}^{\infty} h(y_n)\mathbb{1}_{(Y=y_n)} = \sum_{n=1}^{\infty} \frac{1}{\mathbb{P}(Y = y_n)}\mathbb{E}[X\mathbb{1}_{(Y=y_n)}]\mathbb{1}_{(Y=y_n)}. \quad [2.1]$$

◇

EXAMPLE 2.6.– *The number of students arriving at a university restaurant for a meal is a random variable, M, which follows a Poisson distribution with parameter λ. It is assumed that p is the probability that a student takes a small spoon. The number* *of small spoons required is a random variable, N. Let us calculate the conditional expectation of N knowing M.*

We have $\mathbb{P}(M = m) = \exp(-\lambda)\frac{\lambda^m}{m!}$ *and*

$$\begin{aligned}\mathbb{E}[N\mathbb{1}_{(M=m)}] &= \sum_{n=0}^{\infty} n\mathbb{1}_{n\leq m}\mathbb{P}(N = n, M = m)\\ &= \sum_{n=0}^{m} n\mathbb{P}(N = n|M = m)\mathbb{P}(M = m)\\ &= \sum_{n=0}^{m} nC_m^n p^n(1-p)^{m-n}\exp(-\lambda)\frac{\lambda^m}{m!}\\ &= mp\exp(-\lambda)\frac{\lambda^m}{m!}.\end{aligned}$$

Thus,

$$\mathbb{E}[N|M] = \sum_{m=1}^{\infty} mp\mathbb{1}_{(M=m)} = p\sum_{m=1}^{\infty} m\mathbb{1}_{(M=m)} = Mp.$$

Conditioned by $(M = m)$*, N follows the binomial distribution with parameters* m *and p, with the expectation* mp*; thus, this result is consistent with the intuitive answer.* ◇

REMARK 2.2.– *The conditional expectation is indeed a random variable, despite the term "expectation". As for a conditional probability, the conditioning is denoted by a vertical line. This line does not represent a fraction.* ★

2.2.2. *Properties of conditional expectation*

Theorem 2.1 gives the existence of conditional expectation but does not allow us to carry out explicit calculations except in very specific cases. We thus give a series of properties that are sufficient for carrying out calculation in the majority of cases. Most of these properties are similar to those of conventional expectation, hence the name.

PROPOSITION 2.3.– *Let* $\mathcal{G}$ *be a sub-*σ*-algebra of* $\mathcal{F}$*. We then have* $\mathbb{E}[1|\mathcal{G}] = 1$ *a.s.*

PROOF.– The function $X = 1$ is constant, thus $\mathcal{G}$-measurable (all constants are measurable for all σ-algebras as they are $\{\emptyset, \Omega\}$-measurable). Let $A \in \mathcal{G}$. We then have $\mathbb{E}[X\mathbb{1}_A] = \mathbb{E}[1\mathbb{1}_A]$; thus, we indeed have $\mathbb{E}[X|\mathcal{G}] = \mathbb{E}[1|\mathcal{G}] = 1$ a.s. □

PROPOSITION 2.4.– *Let $X \in L^1(\Omega, \mathcal{F}, \mathbb{P})$ and let $\mathcal{G}$ be a sub-σ-algebra $\mathcal{F}$. We then have*

$$\mathbb{E}[\mathbb{E}[X|\mathcal{G}]] = \mathbb{E}[X].$$

PROOF.– It is known that $\mathbb{E}[X|\mathcal{G}]$ is the single, $\mathcal{G}$-measurable random variable such that for any $A \in \mathcal{G}$ we have $\mathbb{E}\left[\mathbb{E}[X|\mathcal{G}]\mathbb{1}_A\right] = \mathbb{E}\left[X\mathbb{1}_A\right]$. Taking $A = \Omega$, we obtain

$$\mathbb{E}[\mathbb{E}[X|\mathcal{G}]] = \mathbb{E}\left[\mathbb{E}[X|\mathcal{G}]\mathbb{1}_\Omega\right] = \mathbb{E}\left[X\mathbb{1}_\Omega\right] = \mathbb{E}[X].$$

□

PROPOSITION 2.5 (Linearity of conditional expectation).– *Let $X \in L^1(\Omega, \mathcal{F}, \mathbb{P})$ and let $\mathcal{G}$ be a sub-σ-algebra of $\mathcal{F}$. For any $\alpha \in \mathbb{R}$, $Y \in L^1(\Omega, \mathcal{F}, \mathbb{P})$, we have*

$$\mathbb{E}[\alpha X + Y|\mathcal{G}] = \alpha\mathbb{E}[X|\mathcal{G}] + \mathbb{E}[Y|\mathcal{G}] \quad a.s.$$

PROOF.– On the one hand, since the two random variables $\mathbb{E}[X|\mathcal{G}]$ and $\mathbb{E}[Y|\mathcal{G}]$ are $\mathcal{G}$-measurable, their linear combination $\alpha\mathbb{E}[X|\mathcal{G}] + \mathbb{E}[Y|\mathcal{G}]$ is also $\mathcal{G}$-measurable. On the other hand, by fixing $A \in \mathcal{G}$ and using the linearity of the expectation and the characteristic of the conditional expectation (statement 2 in Theorem 2.1), we obtain

$$\begin{aligned}\mathbb{E}\left[(\alpha\mathbb{E}[X|\mathcal{G}] + \mathbb{E}[Y|\mathcal{G}])\,\mathbb{1}_A\right] &= \mathbb{E}\left[\alpha\mathbb{E}[X|\mathcal{G}]\mathbb{1}_A + \mathbb{E}[Y|\mathcal{G}]\mathbb{1}_A\right] \\ &= \alpha\mathbb{E}\left[\mathbb{E}[X|\mathcal{G}]\mathbb{1}_A\right] + \mathbb{E}\left[\mathbb{E}[Y|\mathcal{G}]\mathbb{1}_A\right] \\ &= \alpha\mathbb{E}\left[X\mathbb{1}_A\right] + \mathbb{E}\left[Y\mathbb{1}_A\right] \\ &= \mathbb{E}\left[(\alpha X + Y)\,\mathbb{1}_A\right],\end{aligned}$$

hence the result. □

PROPOSITION 2.6 (Monotonicity in conditional expectation).– *Let X be a random variable in $L^1(\Omega, \mathcal{F}, \mathbb{P})$ and let $\mathcal{G}$ be a sub-σ-algebra of $\mathcal{F}$. Then, for any $Y \in L^1(\Omega, \mathcal{F}, \mathbb{P})$ such that $X \leq Y$ a.s. on a*

$$\mathbb{E}[X|\mathcal{G}] \leq \mathbb{E}[Y|\mathcal{G}] \quad a.s.$$

PROOF.– Let $\epsilon > 0$ and $A_\epsilon = \{\mathbb{E}[X|\mathcal{G}] - \mathbb{E}[Y|\mathcal{G}] \geq \epsilon\}$. We have $A_\epsilon \in \mathcal{G}$ since $\mathbb{E}[X|\mathcal{G}] - \mathbb{E}[Y|\mathcal{G}]$ is $\mathcal{G}$-measurable. Therefore,

$$\mathbb{E}\left[\mathbb{E}[X - Y|\mathcal{G}]\mathbb{1}_{A_\epsilon}\right] = \mathbb{E}\left[(X - Y)\mathbb{1}_{A_\epsilon}\right] \leq 0 \; a.s.$$

Consequently, it follows

$$0 \geq \mathbb{E}\left[\mathbb{E}[X - Y|\mathcal{G}]\mathbb{1}_{A_\epsilon}\right] \geq \epsilon\mathbb{P}(A_\epsilon),$$

which implies that $\mathbb{P}(A_\epsilon) = 0$.

Now, let $A = \{\mathbb{E}[X|\mathcal{G}] - \mathbb{E}[Y|\mathcal{G}] > 0\} = \bigcup_{n\geq 1} A_{1/n}$. Because the sequence $(A_{1/n})_{n\geq 1}$ is increasing, we have

$$\mathbb{P}(A) = \mathbb{P}\Big(\bigcup_{n\geq 1} A_{1/n}\Big) = \lim_{n\to\infty} \mathbb{P}(A_{1/n}) = 0.$$

And so, we have $\mathbb{E}[X|\mathcal{G}] \leq \mathbb{E}[Y|\mathcal{G}]$ *a.s.* □

PROPOSITION 2.7 (Positivity of the conditional expectation).– *Let $X \in L^1(\Omega, \mathcal{F}, \mathbb{P})$ and $\mathcal{G}$ be a sub-σ-algebra of $\mathcal{F}$. If $X \geq 0$ a.s., then we have*

$$\mathbb{E}[X|\mathcal{G}] \geq 0 \;\; a.s.$$

PROOF.– This is a direct application of the previous assertion, taking the random variables X and $Y = 0$. □

THEOREM 2.2 (**Monotone convergence**, conditional expectation version).– *Let $(\Omega, \mathcal{F}, \mathbb{P})$ be a probability space and let $(X_n)_{n\geq 1}$ be a sequence of integrable, real random variables, all defined over this space, such that $0 \leq X_n \leq X_{n+1}$ for any $n \geq 1$, and such that $(X_n)_{n\geq 1}$ converges almost surely towards an integrable random variable X. Thus, for any sub-σ-algebra $\mathcal{G}$ of $\mathcal{F}$, the sequence $(\mathbb{E}[X_n|\mathcal{G}])_{n\geq 1}$ is increasing and converges almost surely to $\mathbb{E}[X|\mathcal{G}]$.*

PROOF.– By writing $Y_n = X - X_n$ and $Z_n = \mathbb{E}[Y_n|\mathcal{G}]$, in order to arrive at the stated result, it is sufficient to demonstrate that the sequence $(Z_n)_{n\geq 1}$ is decreasing, convergent and converges to 0.

Monotonicity and convergence of sequence (Z_n) – since the sequence $(X_n)_{n\geq 1}$ is increasing and positive, the sequence $(Y_n)_{n\geq 1}$ is also positive but decreasing. Using the properties of conditional expectation obtained in Propositions 2.6 and 2.7, (Z_n) is therefore decreasing and $Z_n \geq 0$ for any $n \geq 1$. Consequently, $(Z_n)_{n\geq 1}$ admits an a.s. finite limit, denoted by Z_∞.

Identification of the limit – it remains to be shown that $Z_\infty = 0$ a.s. To do this, and since $Z_\infty \geq 0$ a.s., it is enough to demonstrate that $\mathbb{E}[Z_\infty] = 0$.

The sequence (X_n) converges a.s. to X, $0 \leq X_n \leq X$ for any $n \geq 1$ and $X \in L^1(\Omega, \mathcal{F}, \mathbb{P})$. Thus, using Theorem 1.2 (classical dominated convergence), we obtain $\lim_{n\to\infty} \mathbb{E}[X_n] = \mathbb{E}[X]$. Therefore, it follows that $\lim_{n\to\infty} \mathbb{E}[Y_n] = 0$.

According to Proposition 2.4, we have

$$\mathbb{E}[Y_n] = \mathbb{E}[\mathbb{E}[Y_n|\mathcal{G}]] = \mathbb{E}[Z_n],$$

and we also have $\lim_{n\to\infty} \mathbb{E}[Z_n] = 0$. On the other hand, since $(Z_n)_{n\geq 1}$ is decreasing and $Z_n \leq Z_1$ a.s. for any $n \geq 1$, and further, $\mathbb{E}[Z_1] = \mathbb{E}[Y_1] < \infty$, using Theorem 1.2, we obtain

$$\lim_{n\to\infty} \mathbb{E}[Z_n] = \mathbb{E}[Z_\infty].$$

From this, it can be derived that $\mathbb{E}[Z_\infty] = 0$, and consequently, $Z_\infty = 0$ a.s. □

LEMMA 2.1 (**Fatou's lemma** for conditional expectations).– *Let* $(\Omega, \mathcal{F}, \mathbb{P})$ *be a probability space and let* $(X_n)_{n\geq 1}$ *be a sequence of real random variables, all defined on this space such that* $X_n \geq 0$ *for any* $n \geq 1$. *Therefore, for any sub-*σ*-algebra* $\mathcal{G}$ *of* $\mathcal{F}$, *we have almost surely*

$$\mathbb{E}\left[\liminf_{n\to\infty} X_n|\mathcal{G}\right] \leq \liminf_{n\to\infty} \mathbb{E}[X_n|\mathcal{G}].$$

PROOF.– As the sequence $\left(\inf_{k\geq n} X_k\right)_n$ is increasing, we have

$$\liminf_{n\to\infty} X_n = \sup_n \inf_{k\geq n} X_k = \lim_{n\to\infty} \inf_{k\geq n} X_k \text{ a.s.},$$

and according to Theorem 2.2, we have

$$\mathbb{E}\left[\liminf_{n\to\infty} X_n|\mathcal{G}\right] = \lim_{n\to\infty} \mathbb{E}\left[\inf_{k\geq n} X_k|\mathcal{G}\right]. \qquad [2.2]$$

Further, for any $k \geq n$, we have

$$\mathbb{E}\left[\inf_{k\geq n} X_k|\mathcal{G}\right] \leq \mathbb{E}\left[X_k|\mathcal{G}\right],$$

hence

$$\mathbb{E}\left[\inf_{k\geq n} X_k|\mathcal{G}\right] \leq \inf_{k\geq n} \mathbb{E}\left[X_k|\mathcal{G}\right],$$

and consequently

$$\lim_{n\to\infty} \mathbb{E}\left[\inf_{k\geq n} X_k|\mathcal{G}\right] \leq \lim_{n\to\infty} \inf_{k\geq n} \mathbb{E}\left[X_k|\mathcal{G}\right]. \qquad [2.3]$$

On the other hand, the sequence $\left(\inf_{k\geq n} \mathbb{E}[X_k|\mathcal{G}]\right)_n$ is also increasing; consequently,

$$\liminf_{n\to\infty} \mathbb{E}[X_n|\mathcal{G}] = \sup_n \inf_{k\geq n} \mathbb{E}[X_k|\mathcal{G}] = \lim_{n\to\infty} \inf_{k\geq n} \mathbb{E}[X_k|\mathcal{G}], \; a.s.$$

This, along with the equalities [2.2] and [2.3], makes it possible to obtain the stated result. □

THEOREM 2.3 (**Dominated convergence theorem**, conditional expectation version).– *Let $(\Omega, \mathcal{F}, \mathbb{P})$ be a probability space and let $(X_n)_{n\geq 1}$ be a sequence of real random variables, all defined on this space, such that $|X_n| \leq Z$ for any $n \geq 1$, where Z is an integrable real random variable. If $(X_n)_{n\geq 1}$ converges almost surely to a random variable X, then X is integrable and*

$$\lim_{n\to\infty} \mathbb{E}\left[|X_n - X||\mathcal{G}\right] = 0 \ a.s.$$

In particular, we have

$$\lim_{n\to\infty} \mathbb{E}[X_n|\mathcal{G}] = \mathbb{E}[X|\mathcal{G}] \ a.s.$$

PROOF.– For any $n \geq 1$, since $|X_n| \leq Z$ and since Z is integrable, the random variable X_n is also integrable. By writing

$$Y_n = 2Z - |X_n - X|,$$

because $(X_n)_{n\geq 1}$ converges almost surely to X, we clearly have $(Y_n)_{n\geq 1}$ converges almost surely to $2Z$ and $0 \leq Y_n \leq 2Z$ a.s. Using Lemma 2.1 (Fatou's lemma – the conditional expectation version), we obtain that

$$\mathbb{E}\left[\liminf_{n\to\infty} Y_n|\mathcal{G})\right] \leq \liminf_{n\to\infty} \mathbb{E}[Y_n|\mathcal{G}] \ a.s. \qquad [2.4]$$

Further, we have

$$\mathbb{E}[Y_n|\mathcal{G}] = \mathbb{E}[2Z|\mathcal{G}] - \mathbb{E}[|X_n - X||\mathcal{G}],$$

and consequently,

$$\liminf_{n\to\infty} \mathbb{E}[Y_n|\mathcal{G}] = \mathbb{E}[2Z|\mathcal{G}] - \limsup_{n\to\infty} \mathbb{E}[|X_n - X||\mathcal{G}].$$

This, along with inequality 2.4, makes it possible to obtain

$$\mathbb{E}\left[2Z|\mathcal{G})\right] \leq \mathbb{E}[2Z|\mathcal{G}] - \limsup_{n\to\infty} \mathbb{E}[|X_n - X||\mathcal{G}],$$

which implies that

$$\limsup_{n\to\infty} \mathbb{E}[|X_n - X||\mathcal{G}] \leq 0 \ a.s.$$

Since we also additionally have

$$\limsup_{n\to\infty} \mathbb{E}[|X_n - X||\mathcal{G}] \geq 0 \ a.s.$$

because $|X_n - X| \geq 0$ *a.s.*, it can be derived from this that

$$\limsup_{n\to\infty} \mathbb{E}[|X_n - X||\mathcal{G}] = 0 \ a.s.,$$

and consequently

$$\lim_{n\to\infty} \mathbb{E}[|X_n - X||\mathcal{G}] = 0 \ a.s.$$

Now, since

$$|\mathbb{E}[X_n|\mathcal{G}] - \mathbb{E}[X|\mathcal{G}]| = |\mathbb{E}[X_n - X|\mathcal{G}]| \leq \mathbb{E}[|X_n - X||\mathcal{G}] \to 0 \ a.s.,$$

it can be concluded from this that

$$\lim_{n\to\infty} \mathbb{E}[X_n|\mathcal{G}] = \mathbb{E}[X|\mathcal{G}] \ a.s.,$$

which is indeed the stated result. □

PROPOSITION 2.8 (**The conditional Cauchy–Schwarz inequality**).– *Let X and $Y \in L^2(\Omega, \mathcal{F}, \mathbb{P})$. For any sub-$\sigma$-algebra $\mathcal{G}$ of $\mathcal{F}$,*

$$(\mathbb{E}[XY|\mathcal{G}])^2 \leq \mathbb{E}[X^2|\mathcal{G}]\mathbb{E}[Y^2|\mathcal{G}] \ a.s.$$

PROOF.– For any $x \in \mathbb{R}$, we have

$$\mathbb{E}\left[(X + xY)^2|\mathcal{G}\right] \geq 0 \ a.s.$$

We would now like to switch the $\forall x$ and the *a.s.*, that is, we wish to have a single set with full measure over which the inequality is valid for all x. To do this, we restrict ourselves to rational numbers, for reasons of countability. We have

$$\mathbb{P}\Big(\bigcup_{x\in\mathbb{Q}} \{\mathbb{E}\left[(X + xY)^2|\mathcal{G}\right] < 0\}\Big) \leq \sum_{x\in\mathbb{Q}} \mathbb{P}\Big(\mathbb{E}\left[(X + xY)^2|\mathcal{G}\right] < 0\Big) = 0,$$

and therefore,

$$\mathbb{P}\Big(\bigcap_{x\in\mathbb{Q}} \{\mathbb{E}\left[(X + xY)^2|\mathcal{G}\right] \geq 0\}\Big) = 1.$$

In other words, we have

$$a.s. \, \forall x \in \mathbb{Q}, \ \mathbb{E}\left[(X + xY)^2|\mathcal{G}\right) \geq 0.$$

By linearity, we obtain a.s. for any $x \in \mathbb{Q}$,

$$x^2\mathbb{E}[Y^2|\mathcal{G}] + 2x\mathbb{E}[XY|\mathcal{G}] + \mathbb{E}[X^2|\mathcal{G}] \geq 0.$$

Further, if for any $x \in \mathbb{Q}$, $ax^2 + bx + c \geq 0$, we necessarily have $b^2 - 4ac \leq 0$ (the polynomial cannot have two distinct real roots). By applying this rule with $a = \mathbb{E}[Y^2|\mathcal{G}]$, $b = 2\mathbb{E}[XY|\mathcal{G}]$ and $c = \mathbb{E}[X^2|\mathcal{G}]$, we arrive at the conditional Cauchy–Schwarz inequality. □

PROPOSITION 2.9 (**The conditional Jensen inequality**).– *Let $X \in L^1(\Omega, \mathcal{F}, \mathbb{P})$ and let $\varphi : \mathbb{R} \to \mathbb{R}$ be a convex function such that $\varphi(X) \in L^1(\Omega, \mathcal{F}, \mathbb{P})$. Then,*

$$\varphi(\mathbb{E}[X|\mathcal{F}]) \leq \mathbb{E}\left[\varphi(X)|\mathcal{F})\right] \ a.s.$$

PROOF.– Since φ is convex, if we write

$$L = \{(a,b) \in \mathbb{R}^2; \varphi(x) > ax + b, \forall x \in \mathbb{R}\}$$

then for any $x \in \mathbb{R}$, it is known that

$$\varphi(x) = \sup_{(a,b)\in L\cap\mathbb{Q}^2} \{ax + b\}.$$

We again restrict ourselves to rational values for reasons of countability. Using linearity and increasing conditional expectation, we obtain

$$\begin{aligned}\varphi\left(\mathbb{E}[X|\mathcal{F}]\right) &= \sup_{(a,b)\in L\cap\mathbb{Q}^2} \{a\mathbb{E}[X|\mathcal{F}] + b\} \ a.s.\\ &= \sup_{(a,b)\in L\cap\mathbb{Q}^2} \{\mathbb{E}[aX + b|\mathcal{F}]\} \ a.s.\\ &\leq \mathbb{E}\left[\sup_{(a,b)\in L\cap\mathbb{Q}^2} (aX+b)\Big|\mathcal{F}\right] \ a.s.\\ &= \mathbb{E}[\varphi(X)|\mathcal{F}] \ a.s.,\end{aligned}$$

hence the result. □

COROLLARY 2.1.– *For any $p \geq 1$, the application $X \longmapsto \mathbb{E}[X|\mathcal{G}]$ is a linear and continuous transformation of $L^p(\Omega, \mathcal{F}, \mathbb{P})$ in $L^p(\Omega, \mathcal{G}, \mathbb{P})$.*

PROOF.– Let $p \geq 1$. Since the function $x \longmapsto x^p$ is convex, using the Jensen inequality, we have

$$\Big|\mathbb{E}[X|\mathcal{G}]\Big|^p \leq \mathbb{E}[|X|^p|\mathcal{G}] \ a.s.$$

It follows

$$\mathbb{E}\left[\Big|\mathbb{E}[X\ |\mathcal{G}]\Big|^p\right] \leq \mathbb{E}\left[\mathbb{E}[|X|^p|\mathcal{G}]\right] = \mathbb{E}[|X|^p] \ a.s.$$

We thus obtain

$$||\mathbb{E}[X\,|\mathcal{G}]||_p \leq ||X||_p,$$

and consequently, the linear function $X \longmapsto \mathbb{E}[X|\mathcal{G}]$ is continuous[2]. □

THEOREM 2.4.– ***Chain rule:*** *Let $(\Omega, \mathcal{F}, \mathbb{P})$ be a probability space and let $\mathcal{F}_1$ and $\mathcal{F}_2$ be two sub-σ-algebras of $\mathcal{F}$ such that $\mathcal{F}_1 \subset \mathcal{F}_2$. Then,*

$$\mathbb{E}\left[\mathbb{E}\left[X|\mathcal{F}_2\right]|\mathcal{F}_1\right] = \mathbb{E}\left[X|\mathcal{F}_1\right].$$

PROOF.– We write $U = \mathbb{E}[X|\mathcal{F}_2]$ and $Z = \mathbb{E}[X|\mathcal{F}_1]$. It must be demonstrated that $\mathbb{E}[U|\mathcal{F}_1] = Z$. We have Z $\mathcal{F}_1$-measurable since it is a conditional expectation given $\mathcal{F}_1$. Let $A \in \mathcal{F}_1$. Since $\mathcal{F}_1$ is a sub-σ-algebra of $\mathcal{F}_2$, we also have $A \in \mathcal{F}_2$. Since $Z = \mathbb{E}[X|\mathcal{F}_1]$, on the one hand, we have

$$\mathbb{E}[Z\mathbb{1}_A] = \mathbb{E}[X\mathbb{1}_A],$$

because $A \in \mathcal{F}_1$. Since $U = \mathbb{E}[X|\mathcal{F}_2]$, on the other hand,

$$\mathbb{E}[X\mathbb{1}_A] = \mathbb{E}[U\mathbb{1}_A],$$

because $A \in \mathcal{F}_2$. It follows from this that for any $A \in \mathcal{F}_1$, we have

$$\mathbb{E}[Z\mathbb{1}_A] = \mathbb{E}[U\mathbb{1}_A],$$

which indeed implies that $\mathbb{E}[U|\mathcal{F}_1] = Z$. □

PROPOSITION 2.10.– *Let X and Y be two random variables defined on a probability space $(\Omega, \mathcal{F}, \mathbb{P})$ such that X and XY are integrable and let $\mathcal{G}$ be a sub-σ-algebra of $\mathcal{F}$ such that X is $\mathcal{G}$-measurable. Then,*

$$\mathbb{E}[XY|\mathcal{G}] = X\mathbb{E}[Y|\mathcal{G}] \ a.s.$$

PROOF.– The proof is broken down into several steps. It is based on the fact that a measurable function (therefore a random variable) is written as the simple limit of an increasing sequence of step functions.

2 **Characterization of the continuity of a linear function:** let $(E, \|\cdot\|_E)$ and $(F, \|\cdot\|_F)$ be two normed vector spaces and let $L : E \longrightarrow F$ be a linear function. Then, L is continuous if and only if there exists a constant $C > 0$ such that $\|L(x)\|_F \leq C\|x\|_E$, for any $x \in E$.

Step 1. It is assumed that $X = \mathbb{1}_B$ with $B \in \mathcal{G}$. The two statements of Theorem 2.1 are verified. We have $\mathbb{1}_B \mathbb{E}[Y|\mathcal{G}]$ is $\mathcal{G}$-measurable. Then, for any $A \in \mathcal{G}$, we have

$$\begin{aligned}\mathbb{E}\left[\mathbb{1}_B \mathbb{E}[Y|\mathcal{G}]\mathbb{1}_A\right] &= \mathbb{E}\left[\mathbb{E}[Y|\mathcal{G}]\mathbb{1}_{A\cap B}\right] \\ &= \mathbb{E}\left[Y\mathbb{1}_{A\cap B}\right] \;\; car\; A\cap B \in \mathcal{G} \\ &= \mathbb{E}\left[(\mathbb{1}_B Y)\mathbb{1}_A\right] \;\; a.s.\end{aligned}$$

and therefore

$$\mathbb{1}_B \mathbb{E}[Y|\mathcal{G}] = \mathbb{E}[\mathbb{1}_B Y|\mathcal{G}] \;\; a.s.$$

The result is, thus, true for $\mathcal{G}$-measurable indicator functions.

Step 2. It is assumed that $X = \sum_{i=1}^{n} \alpha_i \mathbb{1}_{B_i}$ with $\alpha_i \in \mathbb{R}$ and $B_i \in \mathcal{G}$. Using the linearity of the conditional expectation, it is possible to write

$$\mathbb{E}[XY|\mathcal{G}] = \mathbb{E}\left[Y\sum_{i=1}^{n}\alpha_i \mathbb{1}_{B_i}|\mathcal{G}\right] = \sum_{i=1}^{n}\alpha_i \mathbb{E}[Y\mathbb{1}_{B_i}|\mathcal{G}] \;\; a.s.$$

According to step 1, we have $\mathbb{E}[Y\mathbb{1}_{B_i}|\mathcal{G}] = \mathbb{1}_{B_i}\mathbb{E}[Y|\mathcal{G}]$ $a.s.$ Therefore,

$$\mathbb{E}[XY|\mathcal{G}] = \sum_{i=1}^{n}\alpha_i \mathbb{1}_{B_i}\mathbb{E}[Y|\mathcal{G}] = X\mathbb{E}[Y|\mathcal{G}] \;\; a.s.,$$

and the result is, therefore, true for $\mathcal{G}$-measurable simple functions.

Step 3. It is assumed that $X, Y \geq 0$. There thus exists an increase in the sequence of random variables $(X_n)_{n\geq 1}$, which almost surely converges to X, and such that $X_n = \sum_{i=1}^{p_n} \alpha_{i,n}\mathbb{1}_{B_{i,n}}$, for any $n \geq 1$. Using step 2, for any $n \geq 1$, we have:

$$\mathbb{E}[X_n Y|\mathcal{G}] = X_n \mathbb{E}[Y|\mathcal{G}] \;\; a.s.$$

On the one hand, the sequence $(X_n\mathbb{E}[Y|\mathcal{G}])$ converges almost surely to $X\mathbb{E}[Y|\mathcal{G}]$. On the other hand, by applying Theorem 2.2 (monotone convergence theorem, conditional version), we obtain that the sequence $(\mathbb{E}[X_nY|\mathcal{G}])$ converges almost surely to $\mathbb{E}[XY|\mathcal{G}]$. It can be derived from this that

$$\mathbb{E}[XY|\mathcal{G}] = X\mathbb{E}[Y|\mathcal{G}] \;\; a.s.,$$

and the result is thus true for positive $\mathcal{G}$-measurable functions, if Y is positive.

Step 4. General case when X and Y have any sign, we decompose them into their positive and negative parts. We write

$$X = X^+ - X^- \; and \; Y = Y^+ - Y^-,$$

with $X^+ = \max\{X, 0\}$ and $X^- = \max\{-X, 0\}$. Using the linearly conditional expectation, the positivity of X^+ and X^- and the result of step 3, we obtain:

$$\begin{aligned}
&\mathbb{E}[XY|\mathcal{G}] = \mathbb{E}[(X^+ - X^-)(Y^+ - Y^-)|\mathcal{G}] \; a.s. \\
&= \mathbb{E}[X^+Y^+|\mathcal{G}] + \mathbb{E}[X^-Y^-|\mathcal{G}] - \mathbb{E}[X^+Y^-|\mathcal{G}] - \mathbb{E}[X^-Y^+|\mathcal{G}] \; a.s. \\
&= X^+\mathbb{E}[Y^+|\mathcal{G}] + X^-\mathbb{E}[Y^-|\mathcal{G}] - X^+\mathbb{E}[Y^-|\mathcal{G}] - X^-\mathbb{E}[Y^+|\mathcal{G}] \; a.s. \\
&= (X^+ - X^-)\mathbb{E}[(Y^+ - Y^-)|\mathcal{G}] \; a.s. \\
&= X\mathbb{E}[Y|\mathcal{G}] \; a.s.
\end{aligned}$$

Hence, the result. □

2.3. Geometric interpretation

We have directly defined conditional expectation via the Radon–Nykodym theorem for integrable variables. A more geometric construction is possible if we restrict ourselves to square integrable random variables. Conditional expectation may then be interpreted as an orthogonal projection.

Let us recall that a **Hilbert space** is a complete vector space H endowed with a scalar product. Let us also recall the definition of the orthogonal projection on a sub-space of a Hilbert space. Let F be a closed sub-space of a Hilbert space H. Then, for any $x \in H$, there exists a unique $y \in F$, written as $y = \pi_F(x)$, which satisfies one of the following equivalent conditions

1) $\forall z \in F, < x - y, z >= 0.$

2) $\forall z \in F, ||x - y||_H \leq ||x - z||_H.$

The vector $y = \pi_F(x)$ is called the **orthogonal projection** of x on F.

Let us now establish the relation between projection and conditional expectation in the Hilbert space $L^2(\Omega, \mathcal{F}, \mathbb{P})$ endowed with the scalar product $< X, Y >= \mathbb{E}[XY]$.

THEOREM 2.5.– *Let $(\Omega, \mathcal{F}, \mathbb{P})$ be a probability space, and*

$$L^2(\mathcal{G}) = L^2(\Omega, \mathcal{G}, \mathbb{P}) = \{Y \; \mathcal{G} - measurable; \mathbb{E}[Y^2] < \infty\}$$

for any sub-σ-algebra $\mathcal{G}$ of $\mathcal{F}$. We then have the following properties:

1) $L^2(\mathcal{G})$ is a Hilbert space for the scalar product

$$< X, Y >= \mathbb{E}[XY].$$

2) For any sub-σ-algebra $\mathcal{H}$ of $\mathcal{G}$, we have $L^2(\mathcal{H})$ is a closed sub-space of $L^2(\mathcal{G})$.

3) The orthogonal projection of X on $L^2(\mathcal{H})$ is $\mathbb{E}[X|\mathcal{H}]$.

PROOF.– The first two statements directly follow from the definition of $L^2(\mathcal{G})$. For the third statement, let $Z \in L^2(\mathcal{H})$. We have

$$\mathbb{E}[ZX|\mathcal{H}] = Z\mathbb{E}[X|\mathcal{H}],$$

and therefore

$$\mathbb{E}\left[Z\mathbb{E}[X|\mathcal{H}]\right] = \mathbb{E}\left[\mathbb{E}[ZX|\mathcal{H}]\right] = \mathbb{E}[ZX],$$

or again

$$\forall Z \in L^2(\mathcal{H}), < X - \mathbb{E}[X|\mathcal{H}], Z >= 0,$$

hence the result. □

REMARK 2.3.– *The fact that the conditional expectation of X given $\mathcal{H}$ is the orthogonal projection of X on $L^2(\mathcal{H})$ makes it possible to interpret $\mathbb{E}[X|\mathcal{H}]$ as the best approximation of X (in the norm L^2) given the information contained in the σ-algebra $\mathcal{H}$.* ★

COROLLARY 2.2.– *Let $X, Y \in L^2(\Omega, \mathcal{F}, \mathbb{P})$. Then, $X - \mathbb{E}[X|Y]$ and Y are non-correlated.*

PROOF.– Using the definition of the orthogonal projection, we have $\mathbb{E}[(X - \mathbb{E}[X|Y])Y] = 0$ and $\mathbb{E}[X - \mathbb{E}[X|Y]] = 0$; therefore, $\text{Cov}(X - \mathbb{E}[X|Y], Y) = 0$. □

2.4. Conditional expectation and independence

To conclude this chapter, let us now explore the relations between conditional expectation and independence.

PROPOSITION 2.11.– *Let X be an integrable random variable and $\mathcal{G}$ be a σ-algebra such that X is independent of $\mathcal{G}$. Then,*

$$\mathbb{E}\left[X|\mathcal{G}\right] = \mathbb{E}[X].$$

In particular, if Y is another random variable independent of X, then $\mathbb{E}[X|Y] = \mathbb{E}[X]$ a.s.

PROOF.– On the one hand, $\mathbb{E}[X]$ is a constant; therefore, it is $\mathcal{G}$-measurable. On the other hand, for any $A \in \mathcal{G}$, because X is independent of $\mathcal{G}$, we have

$$\mathbb{E}[X\mathbb{1}_A] = \mathbb{E}[X]\mathbb{E}[\mathbb{1}_A] = \mathbb{E}[\mathbb{E}[X]\mathbb{1}_A],$$

which implies that $\mathbb{E}[X|\mathcal{G}] = \mathbb{E}[X]$. □

COROLLARY 2.3.– *Let $X \in L^1(\Omega, \mathcal{F}, \mathbb{P})$, f be a bounded or positive measurable function or, more generally, a function such that $f(X)$ is integrable and let $\mathcal{G}$ be a sub-σ-algebra of $\mathcal{F}$. If X is independent of $\mathcal{G}$, then*

$$\mathbb{E}[f(X)|\mathcal{G}] = \mathbb{E}[f(X)].$$

PROOF.– If X is independent of $\mathcal{G}$, then $f(X)$ is also independent of $\mathcal{G}$. It is then sufficient to apply the previous proposition to $f(X)$. □

Thus, an independent σ-algebra of X naturally gives no information on X.

EXAMPLE 2.7.– *Simple symmetric random walk: Let $(X_n)_{n\geq 1}$ be a sequence of independent random variables with the same Rademacher distribution with parameter* $1/2$

$$\mathbb{P}(X_1 = 1) = \mathbb{P}(X_1 = -1) = 1/2.$$

Let $S_0 = 0$ and $S_n = \sum_{k=1}^n X_k$ for $n \geq 1$. Let $(\mathcal{F}_n)_{n\in\mathbb{N}}$ be the natural filtration of $S = (S_n)_{n\in\mathbb{N}}$ and $n \geq 0$. Let us calculate $\mathbb{E}[S_{n+1}|\mathcal{F}_n]$ (all the variables are integrable):

$$\begin{aligned}\mathbb{E}[S_{n+1}|\mathcal{F}_n] &= \mathbb{E}[S_n + X_{n+1}|\mathcal{F}_n]\\ &= \mathbb{E}[S_n|\mathcal{F}_n] + \mathbb{E}[X_{n+1}|\mathcal{F}_n] \qquad \textit{by linearity}\\ &= S_n + \mathbb{E}[X_{n+1}],\end{aligned}$$

because S_n is $\mathcal{F}_n$-measurable and X_{n+1} is independent of $\mathcal{F}_n$. We then use the fact that $\mathbb{E}[X_{n+1}] = 0$ to obtain $\mathbb{E}[S_{n+1}|\mathcal{F}_n] = S_n$. ◇

Conditional expectation may be used to characterize the independence.

PROPOSITION 2.12.– *Let X be a real random variable such that $X \in L^1(\Omega, \mathcal{F}, \mathbb{P})$ and let $\mathcal{G}$ be a sub-σ-algebra of $\mathcal{F}$. The following statements are then equivalent.*

1) X and $\mathcal{G}$ are independent;

2) for any bounded, measurable function f, $\mathbb{E}[f(X)|\mathcal{G}]$ is constant;

3) for any t in $\mathbb{R}$, $\mathbb{E}[\exp(itX)|\mathcal{G}]$ is constant.

In this case, $\mathbb{E}[f(X)|\mathcal{G}] = \mathbb{E}[f(X)]$ *and the characteristic function of* X *is:* $\Phi_X(t) = \mathbb{E}[\exp(itX)]$.

PROOF.– The demonstration uses cyclic implications.

1. $\Longrightarrow$ 2. follows from Proposition 2.3.

2. $\Longrightarrow$ 3. is done by decomposing $\exp(itX)$ into its real part and imaginary part.

3. $\Longrightarrow$ 1. because the real and imaginary parts of the complex exponential functions are dense in bounded measurable functions, therefore in indicator functions. □

PROPOSITION 2.13.– *Let* $X, Y \in L^1(\Omega, \mathcal{F}, \mathbb{P})$ *be real random variables,* $\mathcal{G}$ *a sub-*σ*-algebra of* $\mathcal{F}$ *and* h *measurable function such that* $h(X, Y)$ *is integrable. If* X *is independent from* $\mathcal{G}$ *and* Y *is* $\mathcal{G}$*-measurable, then*

$$\mathbb{E}[h(X, Y)|\mathcal{G}] = H(Y),$$

with $H(y) = \mathbb{E}[h(X, y)]$ *for all* $y \in \mathbb{R}^q$.

PROOF.– Suppose first that h is non negative. Let $A \in \mathcal{G}$. One has

$$\begin{aligned}\mathbb{E}[h(X, Y)\mathbb{1}_A] &= \int_\Omega h(X, Y)\mathbb{1}_A d\mathbb{P} \\ &= \int_{\mathbb{R}^d \times \mathbb{R}^q \times \{0,1\}} h(x, y)a\, d\mathbb{P}_{(X,Y,\mathbb{1}_A)}(x, y, a)\end{aligned}$$

using the transfer theorem, where $\mathbb{P}_{(X,Y,\mathbb{1}_A)}$ is the distribution of the tuple $(X, Y, \mathbb{1}_A)$. As X in independent from $\mathcal{G}$ and both Y and $\mathbb{1}_A$ are $\mathcal{G}$-measurable, we obtain that X is independent from $(Y, \mathbb{1}_A)$ and $\mathbb{P}_{(X,Y,\mathbb{1}_A)} = \mathbb{P}_X \otimes \mathbb{P}_{(Y,\mathbb{1}_A)}$. We therefore have

$$\begin{aligned}\mathbb{E}[h(X, Y)\mathbb{1}_A] &= \int_{\mathbb{R}^d \times \mathbb{R}^q \times \{0,1\}} h(x, y)a\, d\mathbb{P}_{(X,Y,\mathbb{1}_A)}(x, y, a) \\ &= \int_{\mathbb{R}^d \times \mathbb{R}^q \times \{0,1\}} h(x, y)a\, d\mathbb{P}_X \otimes \mathbb{P}_{(Y,\mathbb{1}_A)}(x, y, a) \\ &= \int_{\mathbb{R}^d} \Big(\int_{\mathbb{R}^q \times \{0,1\}} h(x, y)a\, d\mathbb{P}_{(Y,\mathbb{1}_A)}(y, a) \Big) d\mathbb{P}_X(x) \\ &= \mathbb{E}\Big(\int_{\mathbb{R}^q \times \{0,1\}} h(X, y)a\, d\mathbb{P}_{(Y,\mathbb{1}_A)}(y, a) \Big) \\ &= \int_{\mathbb{R}^q \times \{0,1\}} \mathbb{E}[h(X, y)]a\, d\mathbb{P}_{(Y,\mathbb{1}_A)}(y, a)\end{aligned}$$

$$= \int_{\mathbb{R}^q \times \{0,1\}} H(y)a \, d\mathbb{P}_{(Y,\mathbb{1}_A)}(y,a)$$

$$= \int_{\Omega} H(Y)\mathbb{1}_A d\mathbb{P}$$

$$= \mathbb{E}[H(Y)\mathbb{1}_A],$$

hence the result, using Fubini's theorem. For a general integrable function, we separate it into the difference of its positive and negative parts, and apply the above result to each part. □

2.5. Exercises

EXERCISE 2.1.– A person has 5 coins: 2 of these have heads on both sides, 1 has tails on both sides, the other 2 are normal, with heads on one side and tails on the other. It is assumed that all the coins are balanced.

1) The person closes their eyes, chooses a coin at random and tosses it. It turns up heads. What is the probability that the other side of the coin is heads?

2) The person again closes their eyes and tosses the same coin again. What is the probability of obtaining heads?

3) The person opens their eyes and finds that they obtained heads. What is the probability that the other side of the coin is also heads?

4) The person then discards this coin and chooses another at random from the remaining coins and tosses it. What is the probability of obtaining heads?

EXERCISE 2.2.– DNA is subject to endogenous and exogenous mutations. Cells have a repairing mechanism, in order to survive, but sometimes the mutation is fixed and is transmitted to the daughter cells. It is assumed that the number of mutations, M, that the DNA undergoes follows a Poisson distribution with the parameter λ and p denotes the probability that a mutation becomes fixed. It is further assumed that mutations are fixed independently of each other.

1) What is the distribution for the number of fixed mutations F given that $M = n$?

2) Determine the distribution for the couple (M, F).

3) Determine the distribution of F, its expectation and its variance.

4) What is the distribution of M given that $F = k$?

5) Calculate the conditional expectation of M given F.

EXERCISE 2.3.– Let X_1 and X_2 be two independent random variables. It is first assumed that X_1 and X_2 are in binomial distributions with the parameters (n_1, p) and (n_2, p), respectively.

1) Determine the distribution of X_1 given $X_1 + X_2 = n$.

2) Calculate $\mathbb{E}[X_1 | X_1 + X_2]$.

It is now assumed that X_1 and X_2 are in Poisson distributions with the parameters λ_1 and λ_2, respectively.

3) Determine the distribution of X_1 given $X_1 + X_2 = n$.

4) Determine $\mathbb{E}[X_1 | X_1 + X_2]$.

EXERCISE 2.4.– Let X be an integrable random variable and let $\mathcal{F}_0 = \{\emptyset, \Omega\}$. Show that $\mathbb{E}[X | \mathcal{F}_0] = \mathbb{E}[X]$.

EXERCISE 2.5.– Let (X_1, Y) and (X_2, Y) be two couples of integrable random variables.

1) If the two couples have the same distribution, show that $\mathbb{E}[X_1 | Y] = \mathbb{E}[X_2 | Y]$ a.s.

2) It is now assumed that X_1 takes the values -1 and 1 with the probabilities $1/2$, $Y = X_1$ and $X_2 = -X_1$. Calculate $\mathbb{E}[X_1 | Y]$ and $\mathbb{E}[X_2 | Y]$. What do you think?

Let $X_1, \ldots, X_n$ be independent random variables with the same distribution. We write $T = X_1 + \cdots + X_n$.

3) What is the value of $\mathbb{E}[T | X_1]$?

4) Calculate $\mathbb{E}[X_1 | T]$.

We consider the rolling of a red die and a blue die. Let X_i be the result of the ith rolling of the red die, and Y_i the result of the ith rolling of the blue die.

5) Show that the couples $(X_1, X_1 + X_2)$ and $(Y_1, Y_1 + Y_2)$ have the same distribution.

6) Show that $\sigma(X_1 + X_2) \neq \sigma(Y_1 + Y_2)$.

7) Is $\mathbb{E}[X_1 | X_1 + X_2] = \mathbb{E}[Y_1 | Y_1 + Y_2]$ as ?

EXERCISE 2.6.– Let $(\Omega, \mathcal{F}, \mathbb{P})$ be a probability space, let $\mathcal{G}$ be a sub-σ-algebra of $\mathcal{F}$ and X be a random variable with an integrable square. For any $A \in \mathcal{F}$, we write $\mathbb{P}(A | \mathcal{G}) = \mathbb{E}[\mathbb{1}_A | \mathcal{G}]$. Show that for any $a > 0$,

$$\mathbb{P}(|X| \geq a \,|\mathcal{G}) \leq \frac{1}{a^2} \mathbb{E}[X^2 | \mathcal{G}].$$

EXERCISE 2.7.– Let X be a random variable with a uniform distribution over the set $\{-2,-1,1,2\}$ and let $Y = |X|$.

1) Calculate $\mathbb{E}[X]$.

2) Calculate $\mathbb{E}[X|Y]$. Compare the result obtained with that obtained for the first question.

3) Show that X and Y are not independent.

EXERCISE 2.8.– Let X and Y be two independent random variables, each of which has a Bernoulli distribution with the parameter $\frac{1}{2}$.

1) Determine the distribution of $X + Y$.

2) Calculate $\mathbb{E}[X|X+Y]$ and $\mathbb{E}[Y|X+Y]$.

3) Are the random variables $\mathbb{E}[X|X+Y]$ and $\mathbb{E}[Y|X+Y]$ independent?

EXERCISE 2.9.– Let X and Y be two positive random variables, with integrable squares, such that $\mathbb{E}[X|Y] = Y$ a.s. and $\mathbb{E}[Y|X] = X$ a.s.

1) Show that $\mathbb{E}[X^2] = \mathbb{E}[Y^2] = \mathbb{E}[XY]$.

2) Derive from this that $X = Y$ a.s.

EXERCISE 2.10.– Consider a random vector (X,Y) taking values in $\mathbb{N}^2 \setminus \{(0,0)\}$ whose joint distribution is defined by:

$$\forall (i,j) \in \mathbb{N}^2 \setminus \{(0,0)\}, \quad \mathbb{P}(X = i, Y = j) = \frac{(i+j-1)!}{i!j!3^i 6^j \ln 2}.$$

1) Verify that this is indeed a probability distribution on $\mathbb{N}^2 \setminus \{(0,0)\}$.

2) Calculate $\mathbb{P}(X = i)$ for any $i \in \mathbb{N}$.

3) Calculate $\mathbb{E}[Y \mid X]$.

4) From this, derive the following equality $\mathbb{E}[Y] - \dfrac{1}{5}\mathbb{E}[X] = \dfrac{1}{5ln2}$.

EXERCISE 2.11.– Let $(\Omega, \mathcal{F}, \mathbb{P})$ be a probability space and let $\mathcal{G}$ be a sub-σ-algebra of $\mathcal{F}$. Let X be a random variable with integrable square on $(\Omega, \mathcal{F}, \mathbb{P})$. The conditional variance of X given $\mathcal{G}$ is the random variable

$$\mathbb{V}(X|\mathcal{G}) = \mathbb{E}\left[(X - \mathbb{E}[X|\mathcal{G}])^2 \middle| \mathcal{G}\right].$$

1) Show that $\mathbb{V}(X|\mathcal{G}) = \mathbb{E}[X^2|\mathcal{G}] - \left(\mathbb{E}[X|\mathcal{G}]\right)^2$ almost surely.

2) Show that $\mathbb{V}(X) = \mathbb{E}\left[\mathbb{V}(X|\mathcal{G})\right] + \mathbb{V}\left(\mathbb{E}[X|\mathcal{G}]\right)$, where $\mathbb{V}(X)$ denotes the usual variance of X defined by

$$\mathbb{V}(X) = \mathbb{E}[X^2] - \left(\mathbb{E}[X]\right)^2.$$

3) Derive from this that $\mathbb{V}(\mathbb{E}[X|\mathcal{G})] \leq \mathbb{V}(X)$.

4) Show that $\mathbb{V}(\mathbb{E}[X|\mathcal{G}]) = \mathbb{V}(X)$ if and only if X is $\mathcal{G}$-measurable.

EXERCISE 2.12.– Let X be a positive and integrable random variable on $(\Omega, \mathcal{F}, \mathbb{P})$ and let $\mathcal{G}$ be a sub-σ-algebra of $\mathcal{F}$.

1) Show that the event $F = \{\omega \in \Omega; \mathbb{E}[X|\mathcal{G}](\omega) > 0\}$ belongs to $\mathcal{G}$.

2) Show that $(X > 0) \subset F$.

3) Show that F is the smallest set belonging to $\mathcal{G}$ and containing $(X > 0)$.

3

Random Walks

This chapter aims to offer an introduction to an interesting example of stochastic process, simple symmetric walk, with minimum formalism. In most bibliographic references, random walks and the classical problem of the gambler's ruin are treated as applications of the theory of martingales (see Chapter 4) or of Markov chains [BAL 00, BRÉ 99, NOR 98]. An elementary approach is preferred, inspired by [FEL 68], which only requires a knowledge of discrete probabilities and counting. In the next chapter, we will see how certain properties of the random walk can be generalized to the more general class of processes formed by martingales.

Section 3.1 defines the simple symmetric random walk and examines its trajectories. Section 3.2 is dedicated to its long-term behavior. Section 3.3 introduces an initial application in financial mathematics, examining the gambler's ruin problem. Finally, section 3.4 brings together exercises based on the different concepts examined in this chapter, with the solutions given in Chapter 8.

3.1. Trajectories of the random walk

It must be recalled that a *stochastic process* is a sequence of random variables indexed by time (cf. Definition 1.18); it makes it possible to model the temporal evolution of a phenomenon, such as the wealth of a gambler or the value of an asset portfolio, for example. The simple symmetric random walk is the simplest model for a discrete-time process.

Throughout the rest of this chapter, we will work on the probability space $(\Omega, \mathcal{F}, \mathbb{P})$.

3.1.1. *Definition*

Let us begin by defining the simple symmetric random walk.

DEFINITION 3.1.– *A simple symmetric* ***random walk*** *on* $\mathbb{Z}$ *is a sequence* $(S_n)_{n\in\mathbb{N}}$ *of random variables such that*

– $S_0 = x \in \mathbb{Z}$ *is deterministic;*

– $S_{n+1} = S_n + X_{n+1}$ *for any* $n \in \mathbb{N}$,

where $(X_n)_{n\geq 1}$ *is a sequence of independent random variables with the same Rademache distribution with parameter* $1/2$*:*

$$\mathbb{P}(X_1 = -1) = \mathbb{P}(X_1 - 1) = \frac{1}{2}.$$

We use the term *walk* because this process may represent the position of a person who moves in a straight line, with steps of unit length and equal probability of moving forward $(+1)$ or backwards (-1). This walk is said to be *simple* as we only take steps with amplitude 1 and is said to be *symmetric* as each increment can be +1 or -1, with equal probability.

This process may also be used to model the wealth of a gambler who tosses for heads or tails with a balanced coin. Starting from an initial fortune of $S_0 = x > 0$, with each time step, the gambler looses or gains 1€ depending on the toss. The variables X_n thus represent the gain at the n-th game, and S_n is the total fortune of the gambler after the n-th toss.

Based on this definition, we can easily arrive at the average behavior of the walk.

PROPOSITION 3.1.– *For any* $n \in \mathbb{N}$, *we have*

$$\mathbb{E}[S_n] = S_0, \quad \mathit{Var}(S_n) = n.$$

PROOF.– Let us first note that

$$\mathbb{E}[X_1] = (-1) \times \frac{1}{2} + 1 \times \frac{1}{2} = 0,$$

$$\text{Var}(X_1) = \mathbb{E}[X_1^2] = (-1)^2 \times \frac{1}{2} + 1^2 \times \frac{1}{2} = 1,$$

the increments are thus centered and reduced. By the linearity of expectation, we obtain

$$\mathbb{E}[S_n] = \mathbb{E}[S_0 + \sum_{k=1}^{n} X_k] = S_0 + \sum_{k=1}^{n} \mathbb{E}[X_k] = S_0,$$

since S_0 is deterministic. Finally, using the independence of the X_k, we obtain

$$\text{Var}(S_n) = \text{Var}(S_0 + \sum_{k=1}^{n} X_k) = \sum_{k=1}^{n} \text{Var}(X_k) = n,$$

hence, the result. □

Thus, the walk remains constant on average, but its variance becomes greater over time.

3.1.2. *Graphical representation*

As with all discrete-value processes, the random walks have a natural graphical representation. A draw of the walk $(S_n)_{n\in\mathbb{N}}$ over the time interval $[|0, N|]$ is graphically represented by a broken line connecting the points $(0, S_0)$, $(1, S_1)$, ...,(N, S_N). Thus, the x-axis represents time and the y-axis the value of the walk.

EXAMPLE 3.1.– *The trajectory corresponding to the draw* $S_0 = 0$, $X_1 = 1$, $X_2 = 1$, $X_3 = -1$, $X_4 = 1$, $X_5 = -1$ *is given in Figure 3.1.* ◇

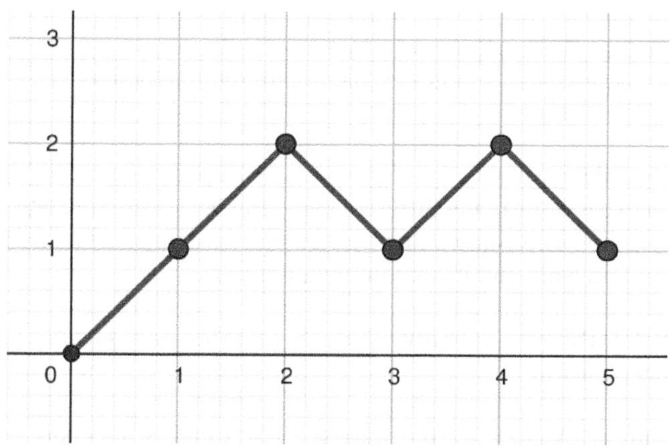

Figure 3.1. *Graphical representation of a trajectory of a random walk between* 0 *and* 5. *For a color version of this figure, see www.iste.co.uk/desaporta/martingales.zip*

The random walk is a discrete-time process; therefore, its graphical representation must be made up of non-connected points (n, S_n). For instance, $S_{1/2}$ is not defined. However, the points are connected for easier reading of the graphs.

We will now look at the graphical representations that may correspond to draws of a random walk.

DEFINITION 3.2.– *Let $n > m$ be two natural integers and let $(a, b) \in \mathbb{Z}^2$ be two integers. The graphical representation of a draw of the random walk, such that $S_m = a$ and $S_n = b$, is called a* **path** *from (m, a) to (n, b).*

EXAMPLE 3.2.– *Figure 3.2 presents two paths from $(1, 1)$ to $(5, 3)$.* ◇

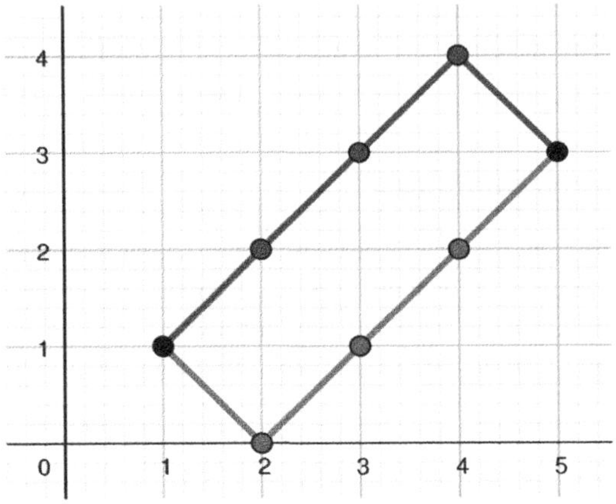

Figure 3.2. *Two paths from $(1, 1)$ to $(5, 3)$. For a color version of this figure, see www.iste.co.uk/desaporta/martingales.zip*

It is now possible to ask whether there is always a path connecting any two points, and, if this path exists, how many different paths can be constructed that connect these two points. It can be easily shown that certain paths are impossible, by looking at the parity of the random walk and its maximum amplitude.

PROPOSITION 3.2 (Parity).– *For any $n \in \mathbb{N}$, S_{2n} has the same parity as S_0 and S_{2n+1} has the opposite parity to S_0.*

PROOF.– For any n, X_{n+1} tales the values -1 or 1 and is thus odd. Since $S_{n+1} = S_n + X_{n+1}$, the walk changes parity with each time step. □

From this, it can be immediately derived that the paths must respect the alternations in parity.

COROLLARY 3.1.– *If $n - m$ and $b - a$ do not have the same parity, then no path exists from (m, a) to (n, b).*

PROOF.– It is assumed without loss of generality that $n - m$ is even and that there exists a path from (m, a) to (n, b). Then, $S_m = a$ and $S_n = b$ necessarily have the same parity; therefore, $S_n - S_m = b - a$ is even. □

EXAMPLE 3.3.– *There is no path from* $(0,0)$ *to* $(2,1)$. ◊

PROPOSITION 3.3 (Amplitude).– *For any* $n \in \mathbb{N}$*, we have* $S_0 - n \leq S_n \leq S_0 + n$.

PROOF.– For any n, X_n takes the values -1 or 1; thus, in particular

$$-1 \leq X_n \leq 1.$$

Since $S_n = S_0 + \sum_{k=1}^{n} X_k$, upon summing the inequalities, we indeed obtain

$$S_0 - n \leq S_n \leq S_0 + n,$$

hence, the result. □

COROLLARY 3.2.– *If* $n - m < |b - a|$*, then no path exists from* (m,a) *to* (n,b).

PROOF.– Let us assume that there exists a path from (m,a) to (n,b). Then, $S_m = a$ and $S_n = b$. In particular, $S_n = S_m + \sum_{k=m+1}^{n} X_k$; hence,

$$\begin{aligned} & S_m - (n-m) \leq S_n \leq S_m + (n-m) \\ \Rightarrow\ & a - (n-m) \leq b \leq a + (n-m) \\ \Rightarrow\ & -(n-m) \leq b - a \leq n - m. \end{aligned}$$

We thus necessarily have $|b - a| \leq n - m$. □

EXAMPLE 3.4.– *There is no path from* $(0,0)$ *to* $(2,4)$. ◊

When the parity conditions and amplitude conditions are satisfied, there are always paths connecting two points, and they can be counted explicitly.

PROPOSITION 3.4 (Number of paths).– *If* $|b - a| \leq n - m$ *and* $n - m$ *and* $b - a$ *have the same parity, then the number of paths from* (m,a) *to* (n,b) *is the number of combinations* $C_{n-m}^{\frac{n-m}{2}+\frac{b-a}{2}}$.

PROOF.– A path from (m,a) to (n,b) corresponds to $n-m$ steps on the random walk, starting from a and arriving at b. It is assumed without loss of generality that $b > a$. It is then possible to break down the steps on the walk into x steps upwards (draws of $+1$) and y steps downwards (draws of -1). We must have a total number of steps $x + y = n - m$, and a difference between the ordinates of $x - y = b - a$, which gives the system

$$\begin{cases} x + y = n - m \\ x - y = b - a \end{cases} \Leftrightarrow \begin{cases} x = \frac{n-m}{2} + \frac{b-a}{2} \\ y = \frac{n-m}{2} - \frac{b-a}{2} \end{cases}.$$

To count the total number of paths of length $x + y$ with x ascents and y descents, it is sufficient to choose the locations of the x ascents among the $x + y$ steps. We thus obtain $C_{x+y}^{x} = C_{n-m}^{\frac{n-m}{2}+\frac{b-a}{2}}$ paths. □

3.1.3. *Reflection principle*

We now wish to count the paths that have a specific additional property, such as that of passing through a fixed point. To do this, we use the symmetric properties of the walk.

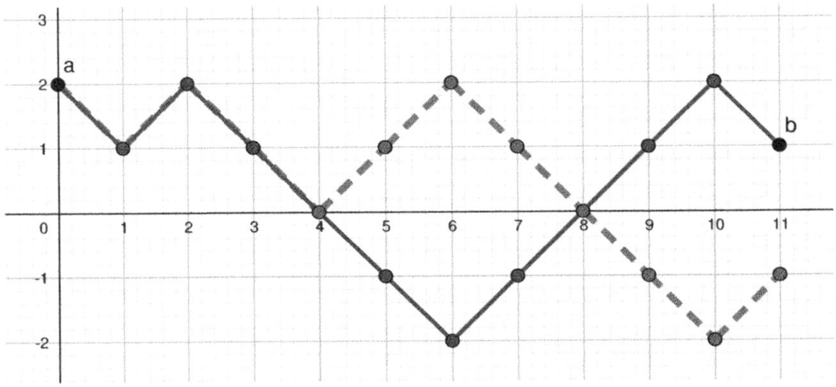

Figure 3.3. *A path from $(0, 2)$ to $(11, 1)$ passing through 0 (the unbroken blue line) and its image (the dotted red line) through symmetry of the first passage through 0*

PROPOSITION 3.5 (Reflection principle).– *Let a and b be two natural non-zero integers, and let n, m be two natural integers such that $n > m$, $|b - a| \leq n - m$ and $b - a$ and $n - m$ have the same parity. Thus, the number of paths from (m, a) to (n, b) going through 0 (i.e. touching the x-axis) is equal to the number of paths from (m, a) to $(n, -b)$.*

PROOF.– The principle is to construct a bijection between the two sets of paths to prove that they have the same cardinal see Figure 3.3. Consider a path $(S_m, \ldots, S_n)$ from (m, a) to (n, b) passing through 0 for the first time at p. We then have $m < p < n$. We construct a path $(\widetilde{S}_m, \ldots, \widetilde{S}_n)$ from (m, a) to $(n, -b)$, retaining the first part of the initial path from (m, a) to $(p, 0)$, and then taking the symmetry of the initial path with respect to the x-axis from $(p, 0)$ to $(n, -b)$:

$$\widetilde{S}_k = \begin{cases} S_k & \text{if } m \leq k \leq p, \\ -S_k & \text{if } p \leq k \leq n. \end{cases}$$

This function defines a bijection between the two sets of paths. Indeed, if we now take a path $(\widetilde{S}_m, \ldots, \widetilde{S}_n)$ from (m, a) to $(n, -b)$, as it takes steps whose amplitude is 1, since $a > 0$ and $-b < 0$, this path must necessarily pass through 0. Let p be the first instant where the path is at 0. We then construct a path $(S_m, \ldots, S_n)$ from (m, a) to (n, b) passing through 0 writing

$$S_k = \begin{cases} \widetilde{S}_k & \text{if } m \leq k \leq p, \\ -\widetilde{S}_k & \text{if } p \leq k \leq n. \end{cases}$$

There are, thus, the same number of paths of both types. □

An initial application of this property is the following result, called the *ballot theorem*.

THEOREM 3.1.– *During an election with two opposing candidates, A and B, candidate A has obtained a votes and candidate B b votes, with a > b. Thus, the probability that A has obtained the majority (in the broad sense) throughout the counting is*

$$p = 1 - \frac{b}{a+1}.$$

PROOF.– With all the counts being equiprobable, p is obtained as the ratio of the number of counts with A in the lead all the time to that of the total number of counts. A count can be modeled by a random walk (S_n), where S_n is the number of votes by which A is ahead of B after counting the n-th ballot.

There are $a + b$ ballots in total and ultimately, A has a lead of $a - b$ votes over B; thus, the total number of counts is the number of paths from $(0, 0)$ to $(a + b, a - b)$ and has the value C_{a+b}^{a}.

The number of counts with A in the lead all long corresponds to

– the number of paths from $(0, 0)$ to $(a + b, a - b)$, not taking any strictly negative value; in other words, the number of paths from $(0, 0)$ to$(a + b, a - b)$ that do not touch -1;

– the number of paths from $(0, 1)$ to $(a + b, a - b + 1)$ not touching 0, by shifting the ordinates upwards by one unit;

– the total number of paths from $(0, 1)$ to $(a + b, a - b + 1)$ minus the number of paths from $(0, 1)$ to $(a + b, a - b + 1)$ touching 0;

– the total number of paths from $(0, 1)$ to $(a + b, a - b + 1)$ minus the number of paths from $(0, 1)$ to $(a + b, -(a - b + 1))$ by the reflection principle.

It thus has the value $C^a_{a+b} - C^{b-1}_{a+b}$. And so,

$$p = \frac{C^a_{a+b} - C^{b-1}_{a+b}}{C^a_{a+b}} = 1 - \frac{C^{b-1}_{a+b}}{C^a_{a+b}} = 1 - \frac{(a+b)!a!b!}{(b-1)!(a+1)!(a+b)!} = 1 - \frac{b}{a+1},$$

hence the result. □

3.2. Asymptotic behavior

We will now look at the behavior of the walk without a time limit. Among the questions of interest are the questions: does the walk returns to its point of origin, x? And how many times and how long does it take?

Since $(X_n)_{n\geq 1}$ are independent and have the same integrable distribution, it is possible to apply the law of large numbers (Theorem 1.3) to obtain that

$$\frac{S_n}{n} \xrightarrow[n\to\infty]{ps} \mathbb{E}[X_1] = 0,$$

which gives no information on the long-term behavior of S_n.

3.2.1. *The Markov property and stationarity property*

The following two properties are no longer simply properties of trajectories, but are properties of processes. They state a certain form of invariance of the walk over time. The first property is that of the absence of memory. The future position of the walk only depends on its current position and not on the trajectory which led to this position.

PROPOSITION 3.6 (Markov).– *For any $n \in \mathbb{N}$ and for all $(a_0, a_1, \ldots, a_{n+1}) \in \mathbb{Z}^{n+2}$, we have*

$$\begin{aligned}&\mathbb{P}(S_{n+1} = a_{n+1} | S_n = a_n, S_{n-1} = a_{n-1}, \ldots, S_1 = a_1, S_0 = a_0)\\ &= \mathbb{P}(S_{n+1} = a_{n+1} | S_n = a_n).\end{aligned}$$

PROOF.– From the definition of conditional probabilities, we have

$$\begin{aligned}&\mathbb{P}(S_{n+1} = a_{n+1} | S_n = a_n, S_{n-1} = a_{n-1}, \ldots, S_1 = a_1, S_0 = a_0)\\ &= \frac{\mathbb{P}(S_{n+1} = a_{n+1}, S_n = a_n, S_{n-1} = a_{n-1}, \ldots, S_1 = a_1, S_0 = a_0)}{\mathbb{P}(S_n = a_n, S_{n-1} = a_{n-1}, \ldots, S_1 = a_1, S_0 = a_0)}\\ &= \frac{\mathbb{P}(X_{n+1} = a_{n+1} - a_n, S_n = a_n, S_{n-1} = a_{n-1}, \ldots, S_1 = a_1, S_0 = a_0)}{\mathbb{P}(S_n = a_n, S_{n-1} = a_{n-1}, \ldots, S_1 = a_1, S_0 = a_0)},\end{aligned}$$

because $S_{n+1} = S_n + X_{n+1}$. It must now be noted that X_{n+1} is independent of $X_1, \ldots, X_n$, therefore of $S_0, \ldots, S_n$. We then obtain

$$\begin{aligned}&\mathbb{P}(S_{n+1} = a_{n+1} | S_n = a_n, S_{n-1} = a_{n-1}, \ldots, S_1 = a_1, S_0 = a_0)\\&= \frac{\mathbb{P}(X_{n+1} = a_{n+1} - a_n)\mathbb{P}(S_n = a_n, S_{n-1} = a_{n-1}, \ldots, S_1 = a_1, S_0 = a_0)}{\mathbb{P}(S_n = a_n, S_{n-1} = a_{n-1}, \ldots, S_1 = a_1, S_0 = a_0)}\\&= \mathbb{P}(X_{n+1} = a_{n+1} - a_n).\end{aligned}$$

Further, using similar reasoning, we have

$$\begin{aligned}\mathbb{P}(S_{n+1} = a_{n+1} | S_n = a_n) &= \frac{\mathbb{P}(S_{n+1} = a_{n+1}, S_n = a_n)}{\mathbb{P}(S_n = a_n)}\\&= \frac{\mathbb{P}(X_{n+1} = a_{n+1} - a_n, S_n = a_n)}{\mathbb{P}(S_n = a_n)}\\&= \frac{\mathbb{P}(X_{n+1} = a_{n+1} - a_n)\mathbb{P}(S_n = a_n)}{\mathbb{P}(S_n = a_n)}\\&= \mathbb{P}(X_{n+1} = a_{n+1} - a_n),\end{aligned}$$

hence, the result. □

PROPOSITION 3.7.– *For any positive integers $n > m$, $S_n - S_m$ is independent of $(S_0, S_1, \ldots, S_m)$.*

PROOF.– As the (X_n) are independent, we directly have

$$S_n - S_m = \sum_{k=m+1}^{n} X_k,$$

which is independent of $(S_0 = x, S_1 = x + X_1, \ldots, S_m = x + \sum_{k=1}^{m} X_k)$. □

We have a result analogous to the Markov property when we look at two more distant time steps.

COROLLARY 3.3.– *For any positive integers $n > m$ and for all $(a_0, a_1, \ldots, a_m, a_n) \in \mathbb{Z}^{n+2}$, we have*

$$\begin{aligned}&\mathbb{P}(S_n = a_n | S_m = a_m, S_{m-1} = a_{m-1}, \ldots, S_1 = a_1, S_0 = a_0)\\&= \mathbb{P}(S_n = a_n | S_m = a_m).\end{aligned}$$

PROOF.– Using a reasoning analogous to the proof for the Markov property, we have

$$
\begin{aligned}
&\mathbb{P}(S_n = a_n | S_m = a_m, S_{m-1} = a_{m-1}, \dots, S_1 = a_1, S_0 = a_0) \\
&= \frac{\mathbb{P}(S_n = a_n, S_m = a_m, S_{m-1} = a_{m-1}, \dots, S_1 = a_1, S_0 = a_0)}{\mathbb{P}(S_m = a_m, S_{m-1} = a_{m-1}, \dots, S_1 = a_1, S_0 = a_0)} \\
&= \frac{\mathbb{P}(S_n - S_m = a_n - a_m, S_m = a_m, S_{m-1} = a_{m-1}, \dots, S_1 = a_1, S_0 = a_0)}{\mathbb{P}(S_m = a_m, S_{m-1} = a_{m-1}, \dots, S_1 = a_1, S_0 = a_0)} \\
&= \frac{\mathbb{P}(S_n - S_m = a_n - a_m)\mathbb{P}(S_m = a_m, S_{m-1} = a_{m-1}, \dots, S_1 = a_1, S_0 = a_0)}{\mathbb{P}(S_m = a_m, S_{m-1} = a_{m-1}, \dots, S_1 = a_1, S_0 = a_0)} \\
&= \mathbb{P}(S_n - S_m = a_n - a_m) \\
&= \mathbb{P}(S_n = a_n | S_m = a_m),
\end{aligned}
$$

using the property of independent increments. □

The second property, called the stationary property, signifies that it is possible to change the origin of the coordinate system without influencing the future behavior of the walk.

PROPOSITION 3.8 (Stationarity).– *For any positive integers $n > m$ and $(a, b) \in \mathbb{Z}^2$, we have*

$$\mathbb{P}(S_n = b | S_m = a) = \mathbb{P}(S_{n-m} = b - a | S_0 = 0).$$

PROOF.– Using the independence of the increments, we obtain

$$
\begin{aligned}
\mathbb{P}(S_n = b | S_m = a) &= \frac{\mathbb{P}(S_n = b, S_m = a)}{\mathbb{P}(S_m = a)} \\
&= \frac{\mathbb{P}(S_n - S_m = b - a, S_m = a)}{\mathbb{P}(S_m = a)} \\
&= \frac{\mathbb{P}(S_n - S_m = b - a)\mathbb{P}(S_m = a)}{\mathbb{P}(S_m = a)} \\
&= \mathbb{P}(S_n - S_m = b - a) \\
&= \mathbb{P}\Big(\sum_{k=m+1}^{n} X_k = b - a\Big) \\
&= \mathbb{P}\Big(\sum_{k=1}^{n-m} X_k = b - a\Big) \\
&= \mathbb{P}(S_{n-m} = b - a | S_0 = 0),
\end{aligned}
$$

as the (X_n) are independent and with the same distribution. □

3.2.2. *First return to 0*

It is now assumed that the walk starts from 0: $S_0 = 0$ and we wish to know whether the walk will return to 0 almost surely. Let us start by observing that if $S_n = 0$, then n is necessarily even.

PROPOSITION 3.9.– *For any $n \in \mathbb{N}$, we have*

$$\mathbb{P}(S_{2n} = 0) = \frac{1}{4^n} C_{2n}^n.$$

PROOF.– This probability corresponds to the number of paths from $(0, 0)$ to $(2n, 0)$ divided by the total number of paths with length $2n$, since all the paths are equiprobable. We thus directly have $\mathbb{P}(S_{2n} = 0) = \frac{C_{2n}^n}{4^n}$. □

We now look at the first instant of the return to 0. It is denoted by T_0. It is, therefore, the random variable

$$T_0 = \inf\{n \geq 1; S_n = 0\},$$

using the convention that $\inf \emptyset = +\infty$. It is, therefore, a discrete random variable taking values in $\mathbb{N}^* \cup \{+\infty\}$. The distribution of T_0 can be explicitly calculated.

PROPOSITION 3.10.– *For any $n \in \mathbb{N}^*$, we have*

$$\mathbb{P}(T_0 = 2n) = \frac{(2n-2)!}{2^{2n-1} n!(n-1)!}.$$

PROOF.– The event $(T_0 = 2n)$ corresponds to $(S_2 \neq 0, \ldots, S_{2n-2} \neq 0, S_{2n} = 0)$ because then $2n$ is the first time the walk returns to 0. In particular, between the instant 0 and the instant $2n$, the walk does not change in sign and retains the same sign as S_1. Therefore, we have

$$\begin{aligned}\mathbb{P}(T_0 = 2n) = {} & \mathbb{P}(S_1 = 1, S_2 > 0, \ldots, S_{2n-2} > 0, S_{2n} = 0) \\ & + \mathbb{P}(S_1 = -1, S_2 < 0, \ldots, S_{2n-2} < 0, S_{2n} = 0).\end{aligned}$$

These two probabilities are equal by symmetry. Further, if $S_{2n-2} > 0$ and $S_{2n} = 0$, then we necessarily have $S_{2n-1} = 1$. Therefore, it follows

$$\begin{aligned}& \mathbb{P}(T_0 = 2n) \\ & = 2\mathbb{P}(S_1 = 1, S_2 > 0, \ldots, S_{2n-2} > 0, S_{2n-1} = 1, S_{2n} = 0) \\ & = 2\mathbb{P}(S_{2n} = 0 | S_1 = 1, S_2 > 0, \ldots, S_{2n-2} > 0, S_{2n-1} = 1) \\ & \times \mathbb{P}(S_1 = 1, S_2 > 0, \ldots, S_{2n-2} > 0, S_{2n-1} = 1)\end{aligned}$$

$$
\begin{aligned}
&= 2\mathbb{P}(S_{2n} = 0 | S_{2n-1} = 1)\mathbb{P}(S_1 = 1, S_2 > 0, \dots, S_{2n-2} > 0, S_{2n-1} = 1) \\
&= 2\mathbb{P}(S_1 = -1 | S_0 = 0)\mathbb{P}(S_1 = 1, S_2 > 0, \dots, S_{2n-2} > 0, S_{2n-1} = 1) \\
&= \mathbb{P}(S_1 = 1, S_2 > 0, \dots, S_{2n-2} > 0, S_{2n-1} = 1),
\end{aligned}
$$

using the Markov property, the stationary property and then the fact that

$$
\mathbb{P}(S_1 = -1 | S_0 = 0) = \mathbb{P}(S_1 = -1) = \mathbb{P}(X_1 = -1) = \frac{1}{2}.
$$

This last probability corresponds to the number of paths from $(1, 1)$ to $(2n - 1, 1)$ that do not touch 0 divided by the total number of paths of length $2n - 1$. Using the reflection principle, we obtain

$$
\begin{aligned}
\mathbb{P}(T_0 = 2n) &= \frac{C_{2n-2}^{n-1} - C_{2n-2}^{n}}{2^{2n-1}} \\
&= \frac{1}{2^{2n-1}}\left(\frac{(2n-2)!}{(n-1)!(n-1)!} - \frac{(2n-2)!}{n!(n-2)!}\right) \\
&= \frac{1}{2^{2n-1}}\frac{(2n-2)!}{(n-1)!(n-2)!}\left(\frac{1}{n-1} - \frac{1}{n}\right) \\
&= \frac{(2n-2)!}{2^{2n-1}n!(n-1)!},
\end{aligned}
$$

hence, the result. □

We can now state the main result of this section.

THEOREM 3.2.– *The random walk starting from* 0 *returns to* 0 *in finite time, with probability* 1

$$
\mathbb{P}(T_0 < +\infty) = 1.
$$

PROOF.– We have $\mathbb{P}(T_0 < +\infty) = \sum_{n=1}^{\infty} \mathbb{P}(T_0 = 2n)$; we must therefore identify the sum of this series. In order to do this, we differently express the probability of the event $(T_0 = 2n)$ by starting from the expression for the number of paths from $(1, 1)$ to $(2n - 1, 1)$ that do not touch 0, divided by the total number of paths of length $2n - 1$. This number of paths is equal to the difference between the number of paths from $(1, 1)$ to $(2n - 1, 1)$ and the number of paths from $(1, 1)$ to $(2n - 1, -1)$ according to the reflection principle. We thus obtain:

$$
\begin{aligned}
&\mathbb{P}(T_0 = 2n) \\
&= \mathbb{P}(S_1 = 1, S_{2n-1} = 1) - \mathbb{P}(S_1 = 1, S_{2n-1} = -1) \\
&= \mathbb{P}(S_{2n-1} = 1 | S_1 = 1)\mathbb{P}(S_1 = 1) - \mathbb{P}(S_{2n-1} = -1 | S_1 = 1)\mathbb{P}(S_1 = 1)
\end{aligned}
$$

$$= \frac{1}{2}\mathbb{P}(S_{2n-2} = 0|S_0 = 0) - \frac{1}{2}\mathbb{P}(S_{2n-2} = -2|S_0 = 0)$$
$$= \frac{1}{2}\mathbb{P}(S_{2n-2} = 0) - \frac{1}{2}\mathbb{P}(S_{2n-2} = -2),$$

by the stationary property and because $S_0 = 0$. Furthermore, we have

$$\begin{aligned}
&\mathbb{P}(S_{2n} = 0) \\
&= \mathbb{P}(S_{2n} = 0, S_{2n-2} = -2) + \mathbb{P}(S_{2n} = 0, S_{2n-2} = 0) \\
&\quad + \mathbb{P}(S_{2n} = 0, S_{2n-2} = 2) \\
&= 2\mathbb{P}(S_{2n} = 0, S_{2n-2} = -2) + \mathbb{P}(S_{2n} = 0, S_{2n-2} = 0),
\end{aligned}$$

by symmetry. It follows that

$$\begin{aligned}
&\mathbb{P}(S_{2n} = 0) \\
&= 2\mathbb{P}(S_{2n} = 0|S_{2n-2} = -2)\mathbb{P}(S_{2n-2} = -2) \\
&\quad + \mathbb{P}(S_{2n} = 0|S_{2n-2} = 0)\mathbb{P}(S_{2n-2} = 0) \\
&= 2\mathbb{P}(S_2 = 2|S_0 = 0)\mathbb{P}(S_{2n-2} = -2) + \mathbb{P}(S_2 = 0|S_0 = 0)\mathbb{P}(S_{2n-2} = 0) \\
&= 2\frac{1}{4}\mathbb{P}(S_{2n-2} = -2) + \frac{1}{2}\mathbb{P}(S_{2n-2} = 0) \\
&= \frac{1}{2}\mathbb{P}(S_{2n-2} = -2) + \frac{1}{2}\mathbb{P}(S_{2n-2} = 0),
\end{aligned}$$

where we have used the fact that

$$\begin{aligned}
\mathbb{P}(S_2 = 2|S_0 = 0) &= \mathbb{P}(X_1 + X_2 = 0) \\
&= \mathbb{P}(X_1 = -1; X_2 = 1) + \mathbb{P}(X_1 = 1; X_2 = -1) \\
&= \mathbb{P}(X_1 = -1)\mathbb{P}(X_2 = 1) + \mathbb{P}(X_1 = 1)\mathbb{P}(X_2 = -1) \\
&= \frac{1}{2}.
\end{aligned}$$

Therefore, we have $\mathbb{P}(S_{2n-2} = -2) = 2\mathbb{P}(S_{2n} = 0) - \mathbb{P}(S_{2n-2} = 0)$.

Let us now return to T_0. We obtain

$$\begin{aligned}
\mathbb{P}(T_0 = 2n) &= \frac{1}{2}\mathbb{P}(S_{2n-2} = 0) - \frac{1}{2}\mathbb{P}(S_{2n-2} = -2) \\
&= \frac{1}{2}\mathbb{P}(S_{2n-2} = 0) - \mathbb{P}(S_{2n} = 0) + \frac{1}{2}\mathbb{P}(S_{2n-2} = 0) \\
&= \mathbb{P}(S_{2n-2} = 0) - \mathbb{P}(S_{2n} = 0).
\end{aligned}$$

Thus, the probability of the event $(T_0 = +\infty)$ may be written using a telescopic sum:

$$\begin{aligned}\mathbb{P}(T_0 = +\infty) &= 1 - \sum_{n=1}^{\infty} \mathbb{P}(T_0 = 2n)\\ &= 1 - \lim_{N\to\infty} \sum_{n=1}^{N} \mathbb{P}(T_0 = 2n)\\ &= 1 - \lim_{N\to\infty} \sum_{n=1}^{N} (\mathbb{P}(S_{2n-2} = 0) - \mathbb{P}(S_{2n} = 0))\\ &= 1 - \lim_{N\to\infty} (\mathbb{P}(S_0 = 0) - \mathbb{P}(S_{2N} = 0))\\ &= \lim_{N\to\infty} \mathbb{P}(S_{2N} = 0)\\ &= \lim_{N\to\infty} \frac{1}{4^N} C_{2N}^N.\end{aligned}$$

Using the Stirling formula, we obtain the equivalence

$$\frac{1}{4^N} C_{2N}^N = \frac{1}{4^N} \frac{(2N)!}{N!N!} \sim \frac{1}{4^N} \frac{(2N)^{2N}\sqrt{4\pi N}e^{-2N}}{(N^N\sqrt{2\pi N}e^{-N})^2} = \frac{1}{\sqrt{\pi N}},$$

and we thus have $\mathbb{P}(T_0 = +\infty) = 0$. □

It is possible to use a similar method to obtain the distribution for the time for reaching any other point. Let $a \in \mathbb{Z}^*$. The time taken to reach a is the random variable

$$T_a = \inf\{n \geq 1; S_n = a\}.$$

THEOREM 3.3.– *The random walk starting from 0 arrives at any state $a \in \mathbb{Z}$ in finite time, with the probability 1. More precisely, if $a \neq 0$, we have $\mathbb{P}(T_a = n) = 0$ if n and a do not have the same parity or if $|a| > n$, and if not*

$$\mathbb{P}(T_a = n) = \frac{|a|}{n} C_n^{\frac{n+a}{2}}.$$

PROOF.– This result is left as an exercise. It is similar to what was done for T_0. □

In particular, the walk starting from 0 will arrive at any point in $\mathbb{Z}$ with a probability of 1; thus, it is almost surely not bounded.

3.3. The Gambler's ruin

To conclude this chapter and as an introduction to problems in financial mathematics, let us look at a problem of ruin. A gambler has an initial capital x.

They toss for heads or tails using a balanced coin and win 1 if they obtain tails and loses 1 if they obtain heads. The gambler has a fixed objective of a fortune $a \geq x$ and a baseline for losing, $b \leq x$. They play until their fortune reaches a or b.

The gambler's fortune is modeled using a random walk $(S_n)_{n\in\mathbb{N}}$, with $S_0 = x$ and $S_n = S_0 + \sum_{k=1}^{n} X_k$, where X_k represents the winnings of the k-th toss.

It is known that arriving at a or b, starting from x is equivalent to arriving at $a - x$ or $b - x$ starting from 0. The game then stops almost surely at the end of a finite time.

We use p_x to denote the probability of ruin, starting from an initial capital of x, and R_x to denote the event *ruined after starting from* x, that is,

$$p_x = \mathbb{P}(T_b < T_a) = \mathbb{P}(R_x).$$

It can be directly noted that $p_a = 0$ and $p_b = 1$ because in both these situations the game does not start; ruin is impossible in the first case and is certain in the second.

We will now obtain a recurrence over p_x. If $b < x < a$, then using the formula for total probabilities, we have

$$\begin{aligned} p_x &= \mathbb{P}(R_x) \\ &= \mathbb{P}(R_x, |S_1 = x+1)\mathbb{P}(S_1 = x+1) + \mathbb{P}(R_x|S_1 = x-1)\mathbb{P}(S_1 = x-1) \\ &= \mathbb{P}(R_{x+1})\frac{1}{2} + \mathbb{P}(R_{x-1})\frac{1}{2}, \end{aligned}$$

by using stationarity. We thus obtain that $(p_x)_{a<x<b}$ is a second-order linear recurrence sequence[1], with the characteristic polynomial $X^2 - 2X + 1 = (X-1)^2$ Therefore, $p_x = \alpha + \beta x$. α and β are identified with the extremal conditions $p_a = 0$ and $p_b = 1$

$$\begin{cases} p_a = 0 = \alpha + \beta a \\ p_b = 1 = \alpha + \beta b \end{cases} \Leftrightarrow \begin{cases} \alpha = \frac{a}{a-b}, \\ \beta = -\frac{1}{a-b}. \end{cases}$$

1 A sequence $(u_n)_{n\in\mathbb{N}}$ is said to be **second-order linear recurrence** if there exists $(a, b) \in \mathbb{R} \times \mathbb{R}^*$ such that for an integer n, we have

$$u_{n+2} = au_{n+1} + bu_n.$$

The polynomial $X^2 - aX + b$ is called the **characteristic polynomial** of the sequence. If the characteristic polynomial has a double root X_0, then there exist two real numbers α and β such that for any integer n,

$$u_n = (\alpha + \beta n)X_0^n.$$

Thus, the probability of ruin, starting from x, is

$$p_x = \frac{a-x}{a-b}.$$

Using similar reasoning, it can be shown that the duration of the game d_x starting from x is the solution to the recurrent linear equation with the right-hand side[2]

$$d_x = \frac{1}{2}(d_{x+1} + d_{x-1}) + 1,$$

with $d_a = d_b = 0$. Its unique solution is $d_x = -ab + x(a+b) - x^2$. In particular, if $x = 0$ and b tend towards $-\infty$, we obtain that $\mathbb{E}[T_a] = +\infty$. It is known that the walk starting from 0 will reach a almost surely, in finite time. However, on average, this time is infinite.

3.4. Exercises

EXERCISE 3.1.– Let (S_n) be a random walk starting from -1. Give the distribution of S_1, S_2, S_3.

EXERCISE 3.2.– Graphically, represent all the possible trajectories of random walk of four steps, starting from 2.

1) How many are there?

2) How many paths end at 0?

3) How many paths end at 2? How many end at 2 and pass through 0? How many end at -2?

4) How many paths end at 3? How many end at an odd number?

---

2 A sequence $(u_n)_{n\in\mathbb{N}}$ is said to be a **second-order linear recurrence with right-hand side** if there exists $(a, b) \in \mathbb{R} \times \mathbb{R}^*$ and a function f, taking values in $\mathbb{R}$ such that for an integer n, we have

$$u_{n+2} = au_{n+1} + bu_n + f(n).$$

The equation $u_{n+2} = au_{n+1} + bu_n$ is then called the associated homogeneous equation.

If $f(t) := X_0^t P(t)$, where X_0 is a double root of the characteristic polynomial of the homogeneous equation and P is a polynomial with real coefficients, then there exist two real numbers α and β and a polynomial Q with the same degree as P, such that for any integer n,

$$u_n = (\alpha + \beta n)X_0^n + n^2 X_0^n Q(n).$$

EXERCISE 3.3.–

1) Graphically, represent all paths from $(0,0)$ to $(3,-1)$. How many are there?

2) Graphically, represent all the paths from $(0,0)$ to $(3,0)$. How many are there?

3) Graphically, represent all the paths from $(0,0)$ to $(3,5)$. How many are there?

EXERCISE 3.4.– 100 people form a queue at a movie hall ticket counter. A seat costs 5 € and 60 have a 5 € note, while the other 40 have 10 € note. How many 5 € notes must the cashier plan to keep so that all the customers are served in the order of arrival with a probability of at least 95%?

EXERCISE 3.5.– Consider a non-symmetric random walk: $S_0 = 0$ and $S_n = \sum_{k=1}^{n} X_k$ with $\mathbb{P}(X_k = 1) = 3/4$ and $\mathbb{P}(X_k = -1) = 1/4$.

1) What values can S_n take?

2) For any $n \geq 1$, we write $Y_n = \frac{X_n+1}{2}$. What is the distribution of Y_1?

3) From this, derive the distribution of S_n and calculate its expectation and its variance.

4) What is the limit of $\frac{S_n}{n}$? What is the limit of S_n?

EXERCISE 3.6.–

1) Show that there are as many paths from 0 to 9 with the fixed length n as there are paths from 0 to 3 of the same length and passing through 6.

2) If a and b are two strictly positive integers, show that there are as many paths from 0 to $a+b$ as there are paths of the same length from 0 to $a-b$ passing through a.

EXERCISE 3.7.– A gambler has 10 € to play with at a slot machine. In each game, the player inserts 1 € into the machine and the machine returns 2 € or nothing, with equal probability.

1) Model the gambler's fortune using a random walk.

Let N be the number of games played until the gambler's ruin.

2) What is the parity of N, that is, the minimal value it can take?

3) Calculate the distribution of N.

EXERCISE 3.8.– Two gamblers, Julie and Thomas, play against each other in a coin-toss game with a balanced coin. Before each toss, the two gamblers lay down 1 € each on the table; if the coin turns up tails, Thomas pockets the 2 €, if it turns up heads, Julie gets the winnings. At the start of the game, Thomas has 10 1 € coins. He does not know Julie's fortune, x. The game ends when one of the two gamblers is ruined.

1) Model Thomas' fortune and its evolution using a random walk (S_n).

2) Model Julie's fortune and its evolution using a random walk (R_n).

3) Calculate $R_0 + S_0$. What is the relationship between R_n and S_n after n rounds?

It is now assumed that Julie's fortune is finite and that the game ends when Thomas is ruined after 26 tosses.

4) Describe the game using a path whose extremities and features are specified.

5) How many paths are there that have the same extremities?

6) How many possible paths correspond to this subset?

4

Martingales

In this chapter, we will define the concept of a "martingale", which is a family of discrete-time stochastic processes that are especially important for applications in financial mathematics that we will see in the following chapters. It will be seen, in particular, that the simple symmetric random walk studied in Chapter 3 is a martingale. This chapter is not an exhaustive course on the theory of martingales. Notably, there is no discussion on convergence, although it is a very important aspect of the general theory of martingales. Readers who wish to have a more comprehensive introduction to the theory of martingales or their use in other fields of application may refer to [DUF 97, WIL 91].

Section 4.1 begins by defining martingales, and section 4.2 looks at the important concept of the transform of a martingale. The Doob decomposition theorem is then introduced in section 4.3. In a certain way, this theorem justifies the universality of martingales. The concept of stopping time is then introduced in section 4.4, and section 4.5 then studies the concept of martingales stopped at a stopping time. Finally, there are a series of exercises in using the concepts explained in this section 4.6.

4.1. Definition

In all that follows, we will work on a filtered probability space $(\Omega, \mathcal{F}, \mathbb{P}, (\mathcal{F}_n)_{n\in\mathbb{N}})$.

DEFINITION 4.1.– *A sequence of random variables $(X_n)_{n\in\mathbb{N}}$ is an $(\mathcal{F}_n)_{n\in\mathbb{N}}$-**martingale** [respectively a **submartingale**, **supermartingale**] if*

1) for any $n \in \mathbb{N}$, X_n is integrable;

2) the sequence $(X_n)_{n\in\mathbb{N}}$ is $(\mathcal{F}_n)_{n\in\mathbb{N}}$-adapted;

3) $\forall n \in \mathbb{N}$, $\mathbb{E}[X_{n+1}|\mathcal{F}_n] = X_n$ [respectively $\mathbb{E}[X_{n+1}|\mathcal{F}_n] \geq X_n$, $\mathbb{E}[X_{n+1}|\mathcal{F}_n] \leq X_n$].

REMARK 4.1.– *To justify that a sequence of random variables $(X_n)_{n\in\mathbb{N}}$ is an $(\mathcal{F}_n)_{n\in\mathbb{N}}$-martingale, it is sufficient to verify statements 1 and 3 in the above definition. Indeed, if for an $n \in \mathbb{N}$, we have $\mathbb{E}[X_{n+1}|\mathcal{F}_n] = X_n$, then X_n is the $\mathcal{F}_n$-measurable. Consequently, the equality in statement 3 implies statement 2. However, in practice, it is often useful to first demonstrate that X_n is $\mathcal{F}_n$-measurable in order to calculate the conditional expectation $\mathbb{E}[X_{n+1}|\mathcal{F}_n]$.* ★

EXAMPLE 4.1.– *The simple symmetric random walk is a martingale with respect to its natural filtration. Indeed, it must be recalled that $(X_n)_{n\geq 1}$ is a sequence of random independent variables and with the same Rademacher distribution with the parameter 1/2*

$$\mathbb{P}(X_1 = 1) = \mathbb{P}(X_1 = -1) = 1/2.$$

Let $S_0 = 0$ and let $S_n = \sum_{k=1}^{n} X_k$ for $n \geq 1$ be the simple symmetric random walk starting from 0. Let $(\mathcal{F}_n)_{n\in\mathbb{N}}$ be the natural filtration of $S = (S_n)_{n\in\mathbb{N}}$ and $n \geq 0$. Then, for any n, $-n \leq S_n \leq n$, therefore S_n is integrable. By construction, (S_n) is adapted with respect to its natural filtration and

$$\mathbb{E}[S_{n+1}|\mathcal{F}_n] = \mathbb{E}[S_n + X_{n+1}|\mathcal{F}_n] = S_n + \mathbb{E}[X_{n+1}] = S_n,$$

because S_n is the $\mathcal{F}_n$-measurable, X_{n+1} is independent of $\mathcal{F}_n$ and $\mathbb{E}[X_{n+1}] = 0$. We thus have a martingale. A martingale corresponds to an equitable game. ◇

EXAMPLE 4.2.– *Let us now study the simple random asymmetric walk. We thus have $(X_n)_{n\geq 1}$, a sequence of independent random variables and with the same Rademacher distribution with parameter $p \neq 1/2$*

$$\mathbb{P}(X_1 = 1) = 1 - \mathbb{P}(X_1 = -1) = p.$$

Let $S_0 = 0$ and $S_n = \sum_{k=1}^{n} X_k$ for $n \geq 1$. Let $(\mathcal{F}_n)_{n\in\mathbb{N}}$ be the natural filtration of $S = (S_n)_{n\in\mathbb{N}}$ and $n \geq 0$. Then, for any n, $-n \leq S_n \leq n$; therefore S_n is integrable. By construction, (S_n) is adapted with respect to its natural filtration and

$$\begin{aligned}\mathbb{E}[S_{n+1}|\mathcal{F}_n] &= \mathbb{E}[S_n + X_{n+1}|\mathcal{F}_n] \\ &= S_n + \mathbb{E}[X_{n+1}] \\ &= S_n + p - (1 - p) \\ &= S_n + 2p - 1.\end{aligned}$$

Therefore, $S = (S_n)_{n\in\mathbb{N}}$ is a supermartingale if $p \leq 1/2$, or a submartingale, if $p > 1/2$. A supermartingale corresponds to a losing game, while a submartingale corresponds to a winning game. ◇

PROPOSITION 4.1.– *A sequence of random variables* $(X_n)_{n\in\mathbb{N}}$ *is an* $(\mathcal{F}_n)_{n\in\mathbb{N}}$*-martingale if and only if*

1) for any $n \in \mathbb{N}$, X_n *is integrable,*

2) $\forall\, n \in \mathbb{N}, \forall\, 0 \leq k \leq n, \mathbb{E}\left[X_n|\mathcal{F}_k\right] = X_k$.

PROOF.– This result is left as an exercise. It can be easily demonstrated by mathematical induction from the definition. □

REMARK 4.2.– *If* $(X_n)_{n\in\mathbb{N}}$ *is an* $(\mathcal{F}_n)_{n\in\mathbb{N}}$*-martingale and if* X *is an integrable and* $\mathcal{F}_0$*-measurable random variable, then the sequence* $(X_n - X)_{n\in\mathbb{N}}$ *is also an* $(\mathcal{F}_n)_{n\in\mathbb{N}}$*-martingale. Thus, at the risk of subtracting* X_0 *from* X_n*, it is possible to restrain our study to the case where* $X_0 = 0$. ★

EXAMPLE 4.3.– *Let* $(X_n)_{n\in\mathbb{N}}$ *be a sequence of random variables such that*

1) for any $n \in \mathbb{N}$, X_n *is integrable and* $\mathbb{E}[X_n] = 0$,

2) for any $n \in \mathbb{N}^*$ *and* $0 \leq p < n$, *the random variables* $X_{n+1} - X_n$ *and* $X_{p+1} - X_p$ *are independent (the sequence* $(X_n)_{n\in\mathbb{N}}$ *is said to be a sequence with* ***independent increments****).*

Therefore, $(X_n)_{n\in\mathbb{N}}$ *is a martingale with respect to its natural filtration. Indeed, by construction, the variables are integrable and the sequence* $(X_n)_{n\in\mathbb{N}}$ *is well-adapted with respect to its natural filtration. Let us calculate the conditional expectation*

$$\begin{aligned}\mathbb{E}[X_{n+1}|\mathcal{F}_n] &= \mathbb{E}[X_{n+1} - X_n + X_n|\mathcal{F}_n] \\ &= \mathbb{E}[X_{n+1} - X_n|\mathcal{F}_n] + X_n,\end{aligned}$$

because X_n *is the* $\mathcal{F}_n$*-measurable. Furthermore,* $X_{n+1} - X_n$ *is independent of* $\mathcal{F}_n$ *according to the hypothesis. We thus obtain*

$$\begin{aligned}\mathbb{E}[X_{n+1}|\mathcal{F}_n] &= \mathbb{E}[X_{n+1} - X_n] + X_n \\ &= \mathbb{E}[X_{n+1}] - \mathbb{E}[X_n] + X_n \\ &= X_n,\end{aligned}$$

since the variables are centered. We thus have a martingale. It may be noted that it would have been sufficient to assume that the X_n *all have the same expectation to obtain the same conclusion.* ◇

EXAMPLE 4.4.– *Let* $(X_n)_{n\in\mathbb{N}}$ *be a sequence of integrable, centered and independent random variables. We consider the sequence sum* $(S_n)_{n\in\mathbb{N}}$ *defined by* $S_n = \sum_{k=0}^{n} X_k$. *Then,* $(S_n)_{n\in\mathbb{N}}$ *is a martingale with respect to its natural filtration* $(\mathcal{F}_n)_{n\in\mathbb{N}}$ *in* $(X_n)_{n\in\mathbb{N}}$.

Indeed, the S_n are integrable and $\mathcal{F}_n$-measurable for any integer n. Let us calculate the conditional expectation

$$\begin{aligned}\mathbb{E}[S_{n+1}|\mathcal{F}_n] &= \mathbb{E}[S_n + X_{n+1}|\mathcal{F}_n] \\ &= S_n + \mathbb{E}[X_{n+1}] \\ &= S_n,\end{aligned}$$

since S_n is the $\mathcal{F}_n$-measurable, X_{n+1} is independent of $\mathcal{F}_n$, and $\mathbb{E}[X_{n+1}] = 0$. ◇

PROPOSITION 4.2.– *Let $(X_n)_{n\in\mathbb{N}}$ be an $(\mathcal{F}_n)_{n\in\mathbb{N}}$-martingale [respectively a submartingale, supermartingale]. Then, the sequence $(\mathbb{E}[X_n])_{n\in\mathbb{N}}$ is constant [respectively increasing, decreasing].*

PROOF.– Let us examine the case where $(X_n)_{n\in\mathbb{N}}$ is an $(\mathcal{F}_n)_{n\in\mathbb{N}}$-submartingale. The other cases are worked in the same way. Let $n \in \mathbb{N}$. We have $\mathbb{E}\left[X_{n+1}|\mathcal{F}_n\right] \geq X_n$ and it follows that

$$\mathbb{E}\left[\mathbb{E}\left[X_{n+1}|\mathcal{F}_n\right]\right] \geq \mathbb{E}\left[X_n\right].$$

Since $\mathbb{E}\left[\mathbb{E}\left[X_{n+1}|\mathcal{F}_n\right]\right] = \mathbb{E}\left[X_{n+1}\right]$, it can be derived from this that $\mathbb{E}\left[X_{n+1}\right] \geq \mathbb{E}\left[X_n\right]$, which implies that the sequence $(\mathbb{E}[X_n])_{n\in\mathbb{N}}$ is an increasing sequence. □

4.2. Martingale transform

THEOREM 4.1.– *Let $(K_n)_{n\in\mathbb{N}}$ be a positive and $(\mathcal{F}_n)_{n\in\mathbb{N}}$ be a predictable process. Consider $(X_n)_{n\in\mathbb{N}}$ an $(\mathcal{F}_n)_{n\in\mathbb{N}}$-martingale [respectively submartingale, supermartingale]. If the process $(K_n)_{n\in\mathbb{N}}$ is bounded, then the process $(K \cdot X_n)_{n\in\mathbb{N}}$, defined by $K \cdot X_0 = X_0$ and for any $n \geq 1$,*

$$K \cdot X_n := \sum_{k=1}^{n} K_k(X_k - X_{k-1}),$$

*is an $(\mathcal{F}_n)_{n\in\mathbb{N}}$-martingale [respectively submartingale, supermartingale]. The process $(K \cdot X_n)_{n\in\mathbb{N}}$ is called a **martingale transform** of (X_n).*

PROOF.– The process $(K \cdot X_n)_{n\in\mathbb{N}}$ is $(\mathcal{F}_n)_{n\in\mathbb{N}}$-adapted, because for any $n \in \mathbb{N}^*$, $K \cdot X_n := \sum_{k=1}^{n} K_k(X_k - X_{k-1})$, is clearly $\mathcal{F}_n$-measurable.

On the contrary, because (K_n) is bounded, there exists $M > 0$ such that

$$\mathbb{E}\left[|K \cdot X_n|\right] \leq M \sum_{k=1}^{n} \mathbb{E}[|X_k - X_{k-1}|] < \infty,$$

because (X_n) is an integrable.

Finally, for any $n \geq 1$, upon simplification we obtain:

$$\begin{aligned}\mathbb{E}\left[K \cdot X_n - K \cdot X_{n-1} | \mathcal{F}_{n-1}\right] &= \mathbb{E}[K_n(X_n - X_{n-1})|\mathcal{F}_{n-1}] \\ &= K_n \mathbb{E}[X_n - X_{n-1}|\mathcal{F}_{n-1}],\end{aligned}$$

where the final equality is due to the fact that (K_n) is $(\mathcal{F}_n)_{n\in\mathbb{N}}$-predictable. Now, when (X_n) is an $(\mathcal{F}_n)_{n\in\mathbb{N}}$ martingale [respectively submartingale, supermartingale], we have

$$\mathbb{E}[X_n - X_{n-1}|\mathcal{F}_{n-1}] = 0 \; [\; resp. \; \geq 0, \; \leq 0\;].$$

The result of the theorem follows from this. □

EXAMPLE 4.5.– *The typical example of a martingale transform is the fortune of a gambler with a predictable betting strategy. Let* (X_n) *be a sequence of independent and identically distributed (i.i.d.) random variables representing the net winnings (positive or negative) of a given game. It is assumed that the game is equitable* $\mathbb{E}[X_1] = 0$*, such that the gambler's fortune* $M_n = M_0 + \sum_{k=1}^{n} X_k$ *is a martingale. Let* (K_n) *be its (predictable) betting strategy. For instance, the gambler begins with an initial stake of* 1*. Each time they win, they continue with the same stake, each time they lose, they double the stakes for the next attempt.*

$$K_{n+1} = K_n \mathbb{1}_{X_n > 0} + 2K_n \mathbb{1}_{X_n < 0}.$$

Therefore, the fortune of the gambler following this strategy

$$M_0 + \sum_{k=1}^{n} K_n X_n = M_0 + \sum_{k=1}^{n} K_n(M_n - M_{n-1})$$

is the transform of the martingale M *using the betting process* K. ◇

The fact that a martingale transform remains a martingale means that if we start with an equitable game, then regardless of the (predictable) betting strategy adopted, it is not possible to increase (or to decrease) the expectation of a win and the game remains equitable.

4.3. The Doob decomposition

It is possible to obtain a martingale starting from any process.

THEOREM 4.2 (Doob decomposition theorem).– *Let* $X = (X_n)_{n\in\mathbb{N}}$ *be a stochastic process that is adapted to the filtration* $(\mathcal{F}_n)_{n\in\mathbb{N}}$ *and integrable. It can then be uniquely decomposed in the form*

$$X_n = X_0 + M_n + A_n$$

with $M_0 = A_0 = 0$*,* $M = (M_n)_{n\in\mathbb{N}}$ *is a martingale, and* $A = (A_n)_{n\in\mathbb{N}}$ *is a predictable process, which is called the* compensator *of* X.

PROOF.– *Existence* We write $A_0 = 0$,

$$A_{n+1} = A_n + \mathbb{E}[X_{n+1} - X_n|\mathcal{F}_n] = \sum_{k=1}^{n} \mathbb{E}[X_{k+1} - X_k|\mathcal{F}_k],$$

$M_0 = 0$ and $M_n = X_n - A_n$ for $n \geq 1$.

We then directly have that (A_n) is predictable and (M_n) is adapted. Since the X_n are integrable, A_n and M_n are also adaptable. Furthermore

$$\begin{aligned}
\mathbb{E}[M_{n+1}|\mathcal{F}_n] &= \mathbb{E}[X_{n+1} - A_{n+1}|\mathcal{F}_n] \\
&= \mathbb{E}[X_{n+1}|\mathcal{F}_n] - A_{n+1} \\
&= \mathbb{E}[X_{n+1}|\mathcal{F}_n] - A_n - \mathbb{E}[X_{n+1} - X_n|\mathcal{F}_n] \\
&= \mathbb{E}[X_{n+1}|\mathcal{F}_n] - A_n - \mathbb{E}[X_{n+1}|\mathcal{F}_n] + X_n \\
&= X_n - A_n \\
&= M_n,
\end{aligned}$$

thus $M = (M_n)_{n\in\mathbb{N}}$ is a martingale.

Unicity Let us assume that there are two such decompositions:

$$X_n = X_0 + M_n + A_n = X_0 + M'_n - A'_n.$$

Then, $A_0 - A'_0 = 0$ and since the processes are predictable

$$\begin{aligned}
A_{n+1} - A'_{n+1} &= \mathbb{E}[A_{n+1} - A'_{n+1}|\mathcal{F}_n] \\
&= \mathbb{E}[M_{n+1} - M'_{n+1}|\mathcal{F}_n] \\
&= M_n - M'_n \\
&= A_n - A'_n
\end{aligned}$$

because we have martingales. Therefore, $A_n = A'_n$ for any n, and consequently, $M_n = M'_n$ for any n. Thus, we do have unicity. □

COROLLARY 4.1.– *In the Doob decomposition of* $X = (X_n)_{n\in\mathbb{N}}$,

– (A_n) *is increasing if and only if* $X = (X_n)_{n\in\mathbb{N}}$ *is a submartingale;*

– (A_n) *is decreasing if and only if* $X = (X_n)_{n\in\mathbb{N}}$ *is a supermartingale;*

– (A_n) *is null if and only if* $X = (X_n)_{n\in\mathbb{N}}$ *is a martingale.*

PROOF.– The final point is a consequence of the first two. By definition, we have $A_{n+1} = A_n + \mathbb{E}[X_{n+1} - X_n|\mathcal{F}_n]$ and $\mathbb{E}[X_{n+1} - X_n|\mathcal{F}_n]$ is positive for a submartingale, negative for a supermartingale. □

EXAMPLE 4.6.– *It has been seen that the simple symmetric walk (S_n) is a martingale. Let us study the Doob decomposition of the process (S_n^2). We have*

$$\begin{aligned}\mathbb{E}[S_{n+1}^2 - S_n^2|\mathcal{F}_n] &= \mathbb{E}[(S_n + X_{n+1})^2 - S_n^2|\mathcal{F}_n] \\ &= \mathbb{E}[S_n^2 + X_{n+1}^2 + 2S_nX_{n+1} - S_n^2|\mathcal{F}_n] \\ &= \mathbb{E}[X_{n+1}^2] + 2S_n\mathbb{E}[X_{n+1}|\mathcal{F}_n] \\ &= 1 + 2S_n\mathbb{E}[X_{n+1}] \\ &= 1.\end{aligned}$$

Therefore, by the preceding demonstration, the Doob decomposition of (S_n^2) is given by

$$A_n = \sum_{k=1}^{n} 1 = n,$$

$$M_n = S_n^2 - n.$$

We automatically obtain that $(S_n^2 - n)$ is a martingale and that (S_n^2) is a submartingale since its compensator is increasing. ◇

4.4. Stopping time

We now introduce the concept of stopping time. Informally, a stopping time corresponds to a random date, thus a date that is unknown in advance, but such that at any instant it is possible to say whether or not the date has passed.

DEFINITION 4.2.– *Let $T : \Omega \longrightarrow \mathbb{N} \cup \{+\infty\}$ be a random variable. It is said that T is an $(\mathcal{F}_n)_{n\in\mathbb{N}}$-**stopping time**, if for any $n \in \mathbb{N}$, $(T \leq n) \in \mathcal{F}_n$.*

EXAMPLE 4.7.– *Let T be a constant positive random variable. Then, T is an $(\mathcal{F}_n)_{n\in\mathbb{N}}$-stopping time, for any filtration $(\mathcal{F}_n)_{n\in\mathbb{N}}$. Indeed, for any natural integer n, we have*

$$(T \leq n) = \begin{cases} \emptyset \; if \; T > n, \\ \Omega \; if \; T \leq n, \end{cases}$$

As any σ-algebra contains the empty set and Ω, we have $(T \leq n) \in \mathcal{F}_n$. ◇

EXAMPLE 4.8.– *Let $(X_n)_{n\in\mathbb{N}}$ be a sequence of real random variables. Let A be a Borel set in $\mathbb{R}$ and consider the random variable*

$$T = \inf\{k \in \mathbb{N}; X_k \in A\},$$

using the convention $\inf \emptyset = +\infty$. *Thus,* T *is a stopping time with respect to the natural filtration of* $(X_n)_{n\in\mathbb{N}}$, *called the* **hitting time** *of the set A. Indeed, for any* $n \in \mathbb{N}$, *we have*

$$(T \leq n) = (X_0 \in A) \cup ... \cup (X_{n-1} \in A) \cup (X_n \in A) \in \mathcal{F}_n$$

because $(\mathcal{F}_n)_{n\in\mathbb{N}}$ *is the natural filtration of* $(X_n)_{n\in\mathbb{N}}$. ◇

PROPOSITION 4.3.– *A random variable* $T : \Omega \longrightarrow \mathbb{N} \cup \{+\infty\}$ *is an* $(\mathcal{F}_n)_{n\in\mathbb{N}}$*-stopping time if and only if, for any* $n \in \mathbb{N}$, $(T = n) \in \mathcal{F}_n$.

PROOF.– Let us assume that T is an $(\mathcal{F}_n)_{n\in\mathbb{N}}$-stopping time. Then, for any $n \in \mathbb{N}^*$, we have: $(T \leq n) \in \mathcal{F}_n$ and $(T \leq n-1) \in \mathcal{F}_{n-1} \subset \mathcal{F}_n$, and consequently,

$$(T = n) = (T \leq n) \cap (T \leq n-1)^c \in \mathcal{F}_n,$$

because a σ-algebra is closed under finite intersection and complements.

It is now assumed that for any $k \in \mathbb{N}$, $(T = k) \in \mathcal{F}_k$. Therefore, for any fixed n, we have

$$(T \leq n) = \bigcup_{k=0}^{n} (T = k) \in \bigcup_{k=0}^{n} \mathcal{F}_k = \mathcal{F}_n,$$

because a filtration is an increasing sequence of σ-algebras. Consequently, T is indeed an $(\mathcal{F}_n)_{n\in\mathbb{N}}$-stopping time □

DEFINITION 4.3.– *Consider an* $(\mathcal{F}_n)_{n\in\mathbb{N}}$*-stopping time,* T. *The set*

$$\mathcal{F}_T = \{A \in \mathcal{F}; \forall n \in \mathbb{N}, A \cap (T \leq n) \in \mathcal{F}_n\}$$

is called the ***σ-algebra of events observable up until time*** T

PROPOSITION 4.4.– *Let* T *be a stopping time and let* $\mathcal{F}_T$ *be the* σ*-algebra of events observable up until* T. *Then,* $\mathcal{F}_T$ *is a* σ*-algebra.*

PROOF.– We have $\emptyset \cap (T \leq n) = \emptyset \in \mathcal{F}_n$ for any n, therefore $\emptyset \in \mathcal{F}_T$.

Let $A \in \mathcal{F}_T$. We then have

$$\begin{aligned} A^c \cap (T \leq n) &= \left(A^c \cap (T \leq n)\right) \cup \left((T \leq n)^c \cap (T \leq n)\right) \\ &= \left(A^c \cup (T \leq n)^c\right) \cap (T \leq n) \\ &= \left(A \cap (T \leq n)\right)^c \cap (T \leq n) \\ &\in \mathcal{F}_n, \end{aligned}$$

for any n. Therefore, $A^c \in \mathcal{F}_T$.

Let $(A_k)_{k\in\mathbb{N}}$ be a sequence of elements of $\mathcal{F}_T$. Then, for any n, we have $\big(\cup_k A_k\big)\cap(T\leq n)=\cup_k\big(A_k\cap(T\leq n)\big)\subset\mathcal{F}_n$; thus, $\cup_k A_k$ is in $\mathcal{F}_T$ which is indeed a σ-algebra. □

REMARK 4.3.– *In the same way that we employed in the demonstration of Proposition 4.3, we can obtain that*

$$\mathcal{F}_T=\{A\in\mathcal{F}_\infty;\forall n\in\mathbb{N},A\cap\{T=n\}\in\mathcal{F}_n\}.$$

★

4.5. Stopped martingales

Let us now see how to combine the concept of a martingale and that of stopping time.

PROPOSITION 4.5.– *Let $(Z_n)_{n\in\mathbb{N}}$ be an $(\mathcal{F}_n)_{n\in\mathbb{N}}$-martingale and let T be an $(\mathcal{F}_n)_{n\in\mathbb{N}}$-stopping time. Consider the sequence $(Z_n^T)_{n\in\mathbb{N}}$ defined as:*

$$Z_n^T=Z_{T\wedge n}=Z_n\mathbb{1}_{\{T>n\}}+Z_T\mathbb{1}_{\{T\leq n\}};\ n\in\mathbb{N}.$$

Then, $(Z_n^T)_{n\in\mathbb{N}}$ is an $(\mathcal{F}_n)_{n\in\mathbb{N}}$-martingale.

PROOF.– We have

$$Z_{T\wedge n}=Z_0+\sum_{l=1}^{T\wedge n}(Z_l-Z_{l-1})=Z_0+\sum_{l\geq 1}\mathbb{1}_{\{l\leq T\wedge n\}}(Z_l-Z_{l-1}).$$

Since $\mathbb{1}_{\{l\leq T\wedge n\}}=\mathbb{1}_{\{l\leq T\}}\mathbb{1}_{\{l\leq n\}}$ and $\mathbb{1}_{\{l\leq n\}}=1$ if and only if $l=1,....,n$, and 0 if $l>n$, it follows that

$$Z_{T\wedge n}=Z_0+\sum_{l=1}^{n}\mathbb{1}_{\{l\leq T\}}(Z_l-Z_{l-1}).$$

On the contrary, for $k\leq n$, we have:

$$\begin{aligned}
&\mathbb{E}\left[Z_{T\wedge n}|\mathcal{F}_k\right]\\
&=\mathbb{E}\left[Z_0|\mathcal{F}_k\right]+\sum_{l=1}^{n}\mathbb{E}\left[\mathbb{1}_{\{l\leq T\}}(Z_l-Z_{l-1})|\mathcal{F}_k\right]\\
&=\mathbb{E}\left[Z_0|\mathcal{F}_k\right]+\sum_{l=1}^{k}\mathbb{E}\left[\mathbb{1}_{\{l\leq T\}}(Z_l-Z_{l-1})|\mathcal{F}_k\right]\\
&=\sum_{l=k+1}^{n}\mathbb{E}\left[\mathbb{1}_{\{l\leq T\}}(Z_l-Z_{l-1})|\mathcal{F}_k\right].
\end{aligned}$$

For the first two terms, because the random variables Z_0 and $\mathbb{1}_{\{l\leq T\}}(Z_l - Z_{l-1})$ are $\mathcal{F}_k$-measurable, for any $l \leq k$, we have

$$\mathbb{E}\left[Z_0|\mathcal{F}_k\right] = Z_0 \ et \ \sum_{l=1}^{k}\mathbb{E}\left[\mathbb{1}_{\{l\leq T\}}(Z_l-Z_{l-1})|\mathcal{F}_k\right] = \sum_{l=1}^{k}\mathbb{1}_{\{l\leq T\}}(Z_l-Z_{l-1}).$$

For the third term, it can be written,

$$\begin{aligned}\sum_{l=k+1}^{n}\mathbb{E}\left[\mathbb{1}_{\{l\leq T\}}(Z_l - Z_{l-1})|\mathcal{F}_k\right] &= \sum_{l=k+1}^{n}\mathbb{E}\left[\mathbb{E}\left[\mathbb{1}_{\{l\leq T\}}(Z_l-Z_{l-1})|\mathcal{F}_{l-1}\right]|\mathcal{F}_k\right]\\ &= \sum_{l=k+1}^{n}\mathbb{E}\left[\mathbb{1}_{\{l\leq T\}}\mathbb{E}\left[(Z_l-Z_{l-1})|\mathcal{F}_{l-1}\right]|\mathcal{F}_k\right]\\ &= 0,\end{aligned}$$

because, as (Z_n) is an $(\mathcal{F}_n)_{n\in\mathbb{N}}$-martingale, we have $\mathbb{E}\left[(Z_l - Z_{l-1})|\mathcal{F}_{l-1}\right] = 0$, for any $l \geq 1$. It follows from this that

$$\mathbb{E}\left[Z_{T\wedge n}|\mathcal{F}_k\right] = Z_0 + \sum_{l=1}^{k}\mathbb{1}_{\{l\leq T\}}(Z_l - Z_{l-1}),$$

and consequently, for any $k \leq n$, we have

$$\mathbb{E}\left[Z_{T\wedge n}|\mathcal{F}_k\right] = Z_{T\wedge k}.$$

Since the $(Z_{T\wedge n})_{n\in\mathbb{N}}$ is also clearly integrable, because $(Z_n)_{n\in\mathbb{N}}$ is integrable, it can be derived from this that $(Z_{T\wedge n})_{n\in\mathbb{N}}$ is indeed an $(\mathcal{F}_n)_{n\in\mathbb{N}}$-martingale. □

PROPOSITION 4.6.– *Let* $(Z_n)_{n\in\mathbb{N}}$ *be an* $(\mathcal{F}_n)_{n\in\mathbb{N}}$*-martingale [respectively submartingale, supermartingale] and let* T *be an* $(\mathcal{F}_n)_{n\in\mathbb{N}}$*-stopping time. The stopped process* $(Z_n^T)_{n\in\mathbb{N}}$ *is therefore an* $(\mathcal{F}_n)_{n\in\mathbb{N}}$*-martingale [respectively submartingale, supermartingale].*

PROOF.– Statements 1 and 3 in Definition 4.1 can be demonstrated in the same way as was done for demonstrating Proposition 4.5. With regard to the statement 2, for any $n \in \mathbb{N}$, we can write

$$Z_{T\wedge n} = Z_0\mathbb{1}_{T=0} + Z_1\mathbb{1}_{T=1} + \ldots + Z_n\mathbb{1}_{T=n} + Z_n\mathbb{1}_{T>n}.$$

Since (Z_n) is $(\mathcal{F}_n)_{n\in\mathbb{N}}$-adapted and T is an $(\mathcal{F}_n)_{n\in\mathbb{N}}$-stopping time, each of the terms in the above sum is $(\mathcal{F}_n)_{n\in\mathbb{N}}$-measurable. It follows from this that $Z_{T\wedge n}$ is the $(\mathcal{F}_n)_{n\in\mathbb{N}}$-measurable and, consequently, the sequence (Z_n^T) is indeed $(\mathcal{F}_n)_{n\in\mathbb{N}}$-adapted. □

We now study the behavior of the random variable X_T defined as: $\forall w \in \Omega$, $X_T(w) = X_{T(w)}(w)$. This is the random variable obtained by replacing the time n in a martingale, with the stopping time T. The variable X_T is called the **terminal variable** of the stopped process $(X_n^T)_{n\geq 0}$, and it is defined when T is almost surely finite.

PROPOSITION 4.7.– *Let $(X_n)_{n\geq 0}$ be an $(\mathcal{F}_n)_{n\in\mathbb{N}}$-adapted process and let T be a stopping time for the same filtration $(\mathcal{F}_n)_{n\in\mathbb{N}}$. Then, the random terminal variable X_T is $\mathcal{F}_T$-measurable.*

PROOF.– Let $n \in \mathbb{N}$ and let A be a Borel set. Then

$$\{X_T \in A\} \cap \{T = n\} = \{X_n \in A\} \cap \{T = n\} \in \mathcal{F}_n$$

because T is an $(\mathcal{F}_n)_{n\in\mathbb{N}}$-stopping time and (X_n) is $(\mathcal{F}_n)_{n\in\mathbb{N}}$-adapted. Therefore, $\{X_T \in A\} \in \mathcal{F}_T$, for any Borel set A. Thus, X_T is indeed $\mathcal{F}_T$ measurable. □

THEOREM 4.3 (**The Doob stopping theorem** for bounded stopping times).– *Let (X_n) be an $(\mathcal{F}_n)$-martingale [respectively a submartingale, supermartingale] and let T be an $(\mathcal{F}_n)$-bounded stopping time. We then have*

$$\mathbb{E}[X_T] = \mathbb{E}[X_0] \text{ [\textit{resp.} } \mathbb{E}[X_T] \geq \mathbb{E}[X_0], \mathbb{E}[X_T] \leq \mathbb{E}[X_0] \text{]}.$$

PROOF.– Since T is bounded, there exists a constant K such that $\mathbb{P}(T \leq K) = 1$. In particular, $T \wedge K = T$; therefore, we have $\mathbb{E}[X_T] = \mathbb{E}[X_{T\wedge K}]$. Furthermore, $(X_n^T)_n$ is a martingale, thus, for any $n \in \mathbb{N}$, we have

$$\mathbb{E}[X_{T\wedge n}] = \mathbb{E}[X_{T\wedge 0}] = \mathbb{E}[X_0].$$

Therefore, by taking the specific case $n = K$, we obtain $\mathbb{E}[X_T] = \mathbb{E}[X_0]$. □

This result may be generalized a stopping time that is only almost surely finite, as long as there is a domination condition to switch the limit and integral.

THEOREM 4.4 (The Doob stopping theorem for dominated martingales).– *Let (X_n) be an $(\mathcal{F}_n)$-martingale [respectively, submartingale, supermartingale] and let T be an $(\mathcal{F}_n)$-stopping time, such that:*

1) $\mathbb{P}(T < \infty) = 1$.

2) *There exists a positive and integrable r.v. Y such that $| X_{T\wedge n} |\leq Y$ a.s, for any $n \in \mathbb{N}$.*

We then have

$$\mathbb{E}[X_T] = \mathbb{E}[X_0] \text{ [\textit{respectively} } \mathbb{E}[X_T] \geq \mathbb{E}[X_0], \mathbb{E}[X_T] \leq \mathbb{E}[X_0] \text{]}.$$

PROOF.– Since T is a.s. finite, we have $X_{T\wedge n} \to X_T$ a.s. when $n \to \infty$. Indeed, let $\Omega' = \{w \in \Omega; T(w) < \infty\}$. We have $\mathbb{P}(\Omega') = 1$ and

$$\forall w \in \Omega', \forall n \geq T(w), X_{T(w)\wedge n}(w) = X_{T(w)}(w).$$

Thus, we obtain

$$\begin{aligned}\mathbb{P}(w\in\Omega; \lim_{n\to\infty} X_{T(w)\wedge n}(w)=X_{T(w)})&=\mathbb{P}(w\in\Omega'; \lim_{n\to\infty} X_{T(w)\wedge n}(w)=X_{T(w)})\\&= \mathbb{P}(w \in \Omega'; X_{T(w)}(w)=X_{T(w)}) = 1.\end{aligned}$$

Using the dominated convergence theorem, we then have

$$\lim_{n\to\infty} \mathbb{E}[X_{T\wedge n}] = \mathbb{E}[X_T].$$

Furthermore, we have $\mathbb{E}[X_{T\wedge n}] = \mathbb{E}[X_0]$, for any integer n, we thus do obtain

$$\mathbb{E}[X_T] = \mathbb{E}[X_0],$$

as expected. □

Once again, these results illustrate how robust the concept of the martingale is: starting from an equitable game, even when stopping at a random moment, we cannot increase (or decrease) the expected gain.

COUNTER-EXAMPLE 4.1.– Consider a coin-toss game. A gambler with 1 € in hand bets on a coin as follows:

– if it turns up tails, they double their fortune;

– if it turns up heads, they lose everything.

Let $X_n \in \{P, F\}$ be the result of the nth toss and let M_n be the gambler's fortune after the nth toss. We have

$$M_{n+1} = \begin{cases} 2M_n & \text{if } X_{n+1} = P \\ 0 & \text{if } X_{n+1} = F, \end{cases}$$

that is, $M_{n+1} = 2M_n \mathbb{1}_{(X_{n+1}=P)}$ and by recurrence, $M_n = 2^n \mathbb{1}_{(X_1=\cdots=X_n=P)}$ because $M_0 = 1$. Therefore, M_n is clearly integrable and this is a process adapted with respect to the filtration generated by (X_n). Let us demonstrate that this is a martingale

$$\mathbb{E}[M_{n+1}|\mathcal{F}_n] = \mathbb{E}[2M_n \mathbb{1}_{(X_{n+1}=P)}|\mathcal{F}_n] = 2M_n \mathbb{E}[\mathbb{1}_{(X_{n+1}=P)}] = M_n.$$

This is indeed a martingale. Let

$$T = \inf\{n; M_n = 0\}$$

be the time of ruin. This is an unbounded stopping time, but $\mathbb{P}(T = \infty) = \lim 2^{-n} = 0$; therefore, it is almost surely finite. We have $\mathbb{E}[M_0] = 1$ but $\mathbb{E}[M_T] = 0$ since $M_T = 0$. Therefore, the stopping theorem cannot be applied (we do not have uniform domination). †

4.6. Exercises

EXERCISE 4.1.– Let $X \in L^1(\Omega)$. For $n \in \mathbb{N}$, we consider the random variable $X_n = \mathbb{E}[X|\mathcal{F}_n]$. Show that the sequence $(X_n)_{n\in\mathbb{N}}$ is an $(\mathcal{F}_n)_{n\in\mathbb{N}}$-martingale.

EXERCISE 4.2.– Let $(X_n)_{n\geq 1}$ be a sequence of positive, independent random variables whose expectation is 1. We write

$$M_0 = 1 \text{ and } M_n = \prod_{i=1}^{n} X_i, \text{ for any } n \geq 1.$$

1) Show that $(M_n)_{n\geq 1}$ is a martingale with respect to the natural filtration $(\mathcal{F}_n)_{n\geq 1}$ of the sequence (X_n).

For $k \geq 1$, we write $a_k = \mathbb{E}[\sqrt{X_k}]$, $N_0 = 1$ and $N_n = \prod_{k=1}^{n} \frac{\sqrt{X_k}}{a_k}$ for $n \geq 1$.

2) Show that for any $k \geq 1$, we have $a_k \in]0, 1]$.

3) Show that the sequence $(N_n)_{n\geq 0}$ is a martingale with respect to $(\mathcal{F}_n)_{n\geq 0}$.

EXERCISE 4.3.– Let $(X_n)_{n\geq 0}$ be a sequence of integrable random variables and adapted with respect to the filtration $(\mathcal{F}_n)_{n\in\mathbb{N}}$. It is assumed that $\mathbb{E}[X_\tau] = \mathbb{E}[X_0]$ for any bounded stopping time τ.

1) Let $n \geq 0$, $A \in \mathcal{F}_n$ and $\tau = n\mathbb{1}_{A^c} + (n+1)\mathbb{1}_A$. Show that τ is an $(\mathcal{F}_n)_{n\geq 0}$-stopping time.

2) Show that for any $A \in \mathcal{F}_n$, we have

$$\mathbb{E}[X_n \mathbb{1}_{A^c} + X_{n+1}\mathbb{1}_A] = \mathbb{E}[X_n].$$

3) Derive from this that (X_n) is an $(\mathcal{F}_n)_{n\geq 0}$-martingale.

EXERCISE 4.4.– Consider $(X_n)_{n\geq 0}$ a sequence of random variables taking values in $\{0, 1\}$, we write:

$$X_0 = 1 \text{ and, } \forall n \in \mathbb{N},\ S_n = \sum_{p=0}^{n} X_p,$$

and it is assumed that for any $n \in \mathbb{N}$, the distribution of X_{n+1} given $(X_0, ..., X_n)$ is the Bernoulli distribution with the parameter $M_n := \frac{S_n}{n+2}$. Finally, consider $\mathcal{F} = (\mathcal{F}_n)_{n\in\mathbb{N}}$ the natural filtration of $(X_n)_{n\in\mathbb{N}}$.

1) Calculate $\mathbb{E}[X_{n+1} \mid \mathcal{F}_n]$.

2) From this, derive that $\mathbb{E}[S_{n+1} \mid \mathcal{F}_n] = \frac{(n+3)S_n}{n+2}$.

3) Show that $(M_n)_{n\in\mathbb{N}}$ is an $\mathcal{F}$-martingale.

EXERCISE 4.5.– Let (X_n) be an $(\mathcal{F}_n)$-martingale and let T be an $(\mathcal{F}_n)$-stopping time, such that $\mathbb{E}[T] < \infty$, and there exists a constant C such that for any integer n,

$$\mathbb{E}[|X_{n+1} - X_n| \mid \mathcal{F}_n] \leq C.$$

1) Show that $\mathbb{P}(T < \infty) = 1$.

We write $W = |X_0| + \sum_{k\geq 0} \mathbb{1}_{\{k<T\}}|X_{k+1} - X_k|$.

2) Show that $\sup_{n\geq 0} |X_{T\wedge n}| \leq W$.

3) Show that $\mathbb{E}[W] < \infty$.

4) Derive from this that $\mathbb{E}[X_T] = \mathbb{E}[X_0]$.

Let $Z_0 = 0$ be the identically zero r.v. and let $(Z_n)_{n\geq 1}$ be a sequence of centered and integrable i.i.d. r.v. Furthermore, let T be a stopping time adapted with respect to the natural filtration $(\mathcal{G}_n)_{n\geq 0}$ of the process $(Z_n)_{n\geq 0}$, and such that $\mathbb{E}[T] < \infty$. Consider the sequence of r.v. defined by $X_n = \sum_{k=0}^{n} Z_k$, for any $n \geq 0$.

5) Show that $\mathbb{E}[X_T] = \mathbb{E}[X_0] = 0$.

Now, let $Z_0 = a$ be a constant r.v. and let $(Z_n)_{n\geq 1}$ be a sequence of integrable i.i.d. random variables that are no longer assumed to be centered. It is also assumed that $\mathbb{E}[Z_1] = a$. Let T still be a stopping time adapted with respect to the natural filtration $(\mathcal{G}_n)_{n\geq 0}$ of the process $(Z_n)_{n\geq 0}$, and such that $\mathbb{E}[T] < \infty$. Again consider a sequence of random variables defined by $X_n = \sum_{k=0}^{n} Z_k$, for any $n \geq 0$.

6) Show that $\mathbb{E}[X_T] = a\mathbb{E}[T]$.

7) It is assumed that $\mathbb{E}[Z_1^2] < \infty$. Show that

$$\mathbb{E}\left[\sum_{k=1}^{T}(Z_k - a)^2\right] = \mathbb{V}(Z_1)\mathbb{E}[T].$$

EXERCISE 4.6.– Let $p \in \left]\frac{1}{2}, 1\right[$ and let $(X_n)_{n\in\mathbb{N}}$ be the sequence of random variables taking values in $\mathbb{Z}$ defined by $X_0 = 0$ and for any $n \geq 1$, $X_n = Y_1 + ... + Y_n$ where (Y_n) is a sequence of i.i.d. random variables and

$$\mathbb{P}(Y_n = 1) = p \;\; et \;\; \mathbb{P}(Y_n = -1) = 1 - p.$$

Let $(\mathcal{F}_n)_{n\geq 0}$ be the natural filtration of (Y_n).

1) Find S, the set of real numbers $\alpha \in \mathbb{R}_+^\star$ such that $\left(\alpha^{X_n}\right)_{n\in\mathbb{N}}$ is an $(\mathcal{F}_n)_{n\geq 0}$-martingale.

For the rest of the exercise, we will take $\alpha \in S$. Let $a, b \geq 0$ and

$$T = \min\{n \in \mathbb{N}; X_n = -a \;\text{ or } X_n = b\}.$$

2) Show that T is an $(\mathcal{F}_n)_{n\geq 0}$-stopping time.

3) Using the strong law of large numbers (Theorem 1.3), show that the sequence of r.v. (X_n) converges a.s. to $+\infty$.

4) Derive from this that $T < \infty$ a.s.

5) Show that for any $t \geq 0$, $\mathbb{E}[\alpha^{X_{T\wedge t}}] = 1$.

6) Derive from this that $\mathbb{E}[\alpha^{X_T}] = 1$.

7) Calculate $\mathbb{P}(X_T = -a)$.

EXERCISE 4.7.– Let $(Z_n)_{n\geq 1}$ be a sequence of independent random variables taking values in $\{-1, 0, 1\}$ such that $Z_0 = 0$ a.s. and

$$\forall n \geq 1, \; \mathbb{P}(Z_n = 1) = \mathbb{P}(Z_n = -1) = \frac{1}{2}.$$

We consider $\mathcal{F} = (\mathcal{F}_n)_{n\in\mathbb{N}}$ the natural filtration of $(Z_n)_{n\in\mathbb{N}}$, and we write

$$S_n = \sum_{i=0}^{n} Z_i,$$

for any $n \geq 0$. Let $a \in \mathbb{N}^*$.

1) Show that $T_a := \inf\{n \geq 0; |S_n| \geq a\}$ is an $\mathcal{F}$-stopping time.

Let us write $T'_a = \inf\{n \in \mathbb{N}; |S_n| = a\}$.

2) Show that $T_a = T'_a$.

Let us fix $\lambda \in]0, \frac{\pi}{2a}[$ and for any $n \in \mathbb{N}$, consider the random variable

$$W_n = \cos(\lambda S_n)(\cos(\lambda))^{-n}.$$

3) Show that $(W_n)_{n\in\mathbb{N}}$ is an $\mathcal{F}$-martingale.
4) Show that for any $n \in \mathbb{N}$, we have

$$\cos(\lambda a)(\cos(\lambda))^{-\min(n,T_a)} \leq W_{\min(n,T_a)}.$$

5) Show that for any $n \in \mathbb{N}$,

$$\mathbb{E}\Big[\cos(\lambda a)(\cos(\lambda))^{-\min(n,T_a)}\Big] \leq 1.$$

EXERCISE 4.8.– Let $(X_n)_{n\geq 0}$ be a sequence of random variables defined on the same probability space. This sequence is said to be *uniformly integrable* if we have

$$\lim_{c\to\infty} \sup_{n\geq 0} \mathbb{E}\Big[\mathbb{1}_{\{|X_n|>c\}}|X_n|\Big] = 0.$$

1) Show that if $(X_n)_{n\geq 0}$ is uniformly integrable, then there exists $M > 0$ such that for any $n \geq 0$, we have $\mathbb{E}[|X_n|] \leq M$.

Let $(X_n)_{n\geq 0}$ be an $(\mathcal{F}_n)_{n\geq 0}$-martingale and let T be an $(\mathcal{F}_n)_{n\geq 0}$-stopping time such that $\mathbb{P}(T < \infty) = 1$.

2) Show that if the sequence $(X_n)_{n\geq 0}$ is uniformly integrable, then $\mathbb{E}[|X_T|] < \infty$ and $\mathbb{E}[X_T] = \mathbb{E}[X_0]$.

EXERCISE 4.9.– Two gamblers, Mounir and Benoîte bet simultaneously on a 1 € coin and the game ends when one of the two is ruined. Mounir initially has a € and Benoîte has b €.

1) What is the average value of Mounir's fortune at the end of the game?

2) What is the probability that Mounir wins Benoîte's fortune?

EXERCISE 4.10.– Let $(\mathcal{F}_n)_{n\geq 0}$ be a filtration and let $(X_n)_{n\geq 0}$ be an $(\mathcal{F}_n)_{n\geq 0}-$martingale.

1) Show that if T is an $(\mathcal{F}_n)_{n\geq 0}-$ finite stopping time, then X_T is an $\mathcal{F}_T$-measurable.

Let S and T be two $(\mathcal{F}_n)_{n\geq 0}-$ stopping times bounded by a constant $c \in \mathbb{N}$ and such that $S \leq T$.

2) Show that $\mathcal{F}_S \subset \mathcal{F}_T$.

Let $A \in \mathcal{F}_S$ and $R = S\mathbb{1}_A + T\mathbb{1}_{A^c}$.

3) Show that R is an $(\mathcal{F}_n)_{n\geq 0}-$ stopping time.

4) From this derive that $\mathbb{E}[X_R] = \mathbb{E}[X_0]$.

5) From this derive that $\mathbb{E}[X_T\mathbb{1}_A] = \mathbb{E}[X_S\mathbb{1}_A]$.

6) From this derive that $\mathbb{E}[X_T|\mathcal{F}_S] = X_S$.

EXERCISE 4.11.– Let $(\Omega, \mathcal{F}, \mathbb{P}, (\mathcal{F}_n)_{n\in\mathbb{N}})$ be a filtered probability space and let $(X_n)_{n\geq 0}$ be a positive $(\mathcal{F}_n)_{n\in\mathbb{N}}$ submartingale.

Let $(K_n)_{n\in\mathbb{N}}$ be a process that is positive, bounded and $(\mathcal{F}_n)_{n\in\mathbb{N}}$- predictable. Consider $((K \cdot X)_n)_{n\in\mathbb{N}}$, the process defined by $(K \cdot X)_0 = 0$ and for any $n \geq 1$,

$$(K \cdot X)_n = \sum_{k=1}^{n} K_k(X_k - X_{k-1}). \qquad (\star)$$

1) Show that $((K \cdot X)_n)_{n\in\mathbb{N}}$ is an $(\mathcal{F}_n)_{n\in\mathbb{N}}$ submartingale.

Let T be an $(\mathcal{F}_n)_{n\in\mathbb{N}}$-stopping time such that there exists $n_0 \in \mathbb{N}^*$; $\forall w \in \Omega, T(w) \leq n_0$.

2) Show that $\mathbb{E}[X_0] \leq \mathbb{E}[X_T]$.

Let $((K \cdot X)_n)_{n\in\mathbb{N}}$ be the sequence of random variables defined by the formula $(\star)$ with $K_k = \mathbb{1}_{\{T\leq k-1\}}$ for any $k \geq 1$.

3) Show that $((K \cdot X)_n)_{n\in\mathbb{N}}$ is an $(\mathcal{F}_n)_{n\in\mathbb{N}}$ submartingale.

4) Derive from this that $\mathbb{E}[X_T] \leq \mathbb{E}(X_{n_0}]$.

Let $\lambda > 0$, $A_n := \{\max_{0\leq k\leq n} X_k \geq \lambda\}$ and T be the function defined on Ω by:

$$T : w \longmapsto T(w) = \inf\{k; , X_k \geq \lambda\} \wedge n.$$

5) Show that T is an $(\mathcal{F}_n)_{n\in\mathbb{N}}$-stopping time.

6) Show that

$$\lambda \mathbb{P}(A_n) \leq \mathbb{E}[X_T \mathbb{1}_{A_n}].$$

7) Derive from this that

$$\mathbb{P}\Big(\max_{0 \leq k \leq n} X_k \geq \lambda \Big) \leq \frac{\mathbb{E}[X_n]}{\lambda}.$$

5

Financial Markets

In this chapter, we begin to discuss financial mathematics properly. We define the basic vocabulary and the objects that will be of use in the following chapters. The main references for this chapter and the following two chapters are [VIN 04, HUL 18, SHR 03, SHR 04]. As stated in the Introduction, most texts focus largely or even exclusively on continuous-time models. However, we will be focusing on discrete-time models.

Section 5.1 introduces the concept of a financial asset, section 5.2 introduces the concept of an investment strategy, and section 5.3 introduces the concept of arbitration. This key concept allows us to establish a relationship with the martingale theory, presented in Chapter 4. Section 5.4 introduces a discrete-time financial market model: the Cox, Ross and Rubinstein model. This model will be used to illustrate all the concepts, defined in a generic manner for any discrete financial markets. The chapter concludes with exercises in section 5.5 and practical work in section 5.6, which presents simulations and portfolio optimization using the Cox, Ross and Rubinstein model.

It is assumed here, and in the following, that Ω has a finite cardinal and that $\mathbb{P}(\omega) > 0$ for any $\omega \in \Omega$. These conditions are not strictly necessary; however, they make it possible to considerably simplify all questions of measurability, integrability and the existence of an optimum.

Let N be a time horizon and let $(\mathcal{F}_n)_{0 \leq n \leq N}$ be a filtration such that $\mathcal{F}_0 = \{\emptyset, \Omega\}$ and $\mathcal{F}_N = \mathcal{F}$.

5.1. Financial assets

Let us consider a financial market comprising of a risk-free asset $(S_n^0)_{0\leq n\leq N}$ and $d \geq 1$ risky assets $(S_n^0)_{0\leq n\leq N}$ for $1 \leq i \leq d$. The price of the risky assets is represented by stochastic processes. The risk-free asset dynamic is

$$S_0^0 = 1,$$

$$S_n^0 = (1+r)^n,\ n \geq 1,$$

where r is the *interest rate* on the market. The risky assets have a random dynamic that is unspecified for the moment. As we need to compare the prices of the assets on different dates, we may sometimes wish to ignore the effect of the depreciation of currency, which is considered to be linked to the risk-free rate of interest.

DEFINITION 5.1.– *The **discounted prices** of assets are the prices divided by the current value of the risk-free asset*

$$\widetilde{S}_n^i = \frac{S_n^i}{S_n^0} = \frac{S_n^i}{(1+r)^n}, \qquad n \geq 0.$$

In mathematical terms, discounting prices is equivalent to considering that the risk-free interest rate is zero.

5.2. Investment strategies

Let us now consider an investor who wishes to invest money in the market. In order to define their investment strategy, they must know, at each instant, the number of shares invested in each asset.

DEFINITION 5.2.– *An **investment strategy** is a process $\Phi = (\Phi_n)_{1\leq n\leq N}$ that corresponds to the quantities $\Phi_n = (\phi_n^0, \phi_n^1, \dots, \phi_n^d) \in \mathbb{R}^{d+1}$ of each asset held by an investor between the instant $n-1$ and the instant n. It is a predictable process: for any $1 \leq n \leq N$, and $0 \leq i \leq d$, ϕ_n^i is the $\mathcal{F}_{n-1}$ measurable.*

An investment strategy must be predictable, since decisions on how to distribute the portfolio between the instants $n-1$ and n can be based only on the information available up to the instant $n-1$. In other words, there is no insider trading and the investor has no information on the future of the market.

Given the current prices of assets and an investment strategy, it is possible to calculate the value of the investor's portfolio at every instant.

DEFINITION 5.3.– *The value, at an instant n, of a portfolio that applies the investment strategy Φ is*

$$V_n(\Phi) = \sum_{i=1}^{d} \phi_n^i S_n^i = < \Phi_n, S_n >,$$

where $< x, y >$ is the scalar product of x and y, and $S_n = (S_n^0, S_n^1, \ldots, S_n^d)$.

The above notation may be ambiguous. Indeed, using our notations, $V_n(\Phi)$ represents the wealth at time n, just after the asset prices have been updated and before the portfolio is redistributed for the next period. We may also wish to consider the wealth at the instant n after the redistribution of the portfolio. In fact, in most cases that we will consider, these two quantities are equal.

DEFINITION 5.4.– *An investment strategy Φ is said to be* **self-financed** *if for any $1 \leq n \leq N-1$, we have*

$$\sum_{i=1}^{d} \phi_n^i S_n^i = \sum_{i=1}^{d} \phi_{n+1}^i S_n^i,$$

or $< \Phi_n, S_n > = < \Phi_{n+1}, S_n >$, using condensed notation.

A self-financed strategy is, therefore, a strategy where at each step, the entire wealth is reinvested without withdrawal or an exogenous infusion of money.

We will now describe self-financing strategies through a specific decomposition of the realized wealth.

PROPOSITION 5.1.– *An investment strategy Φ is self-financed if and only if, for any $1 \leq n \leq N-1$, we have*

$$\widetilde{V}_n(\Phi) = \widetilde{V}_0(\Phi) + \sum_{k=1}^{n} < \Phi_k, (\widetilde{S}_k - \widetilde{S}_{k-1}) > .$$

PROOF.– We proceed using double implication.

Let us first assume that Φ is self-financed. We then have, by definition, $< \Phi_n, S_n > = < \Phi_{n+1}, S_n >$. Upon dividing by S_n^0, we obtain $< \Phi_n, \widetilde{S}_n > = < \Phi_{n+1}, \widetilde{S}_n >$. Therefore, using the self-financing relation, we obtain

$$\begin{aligned}\widetilde{V}_n(\Phi) &= < \Phi_n, \widetilde{S}_n > \\ &= < \Phi_0, \widetilde{S}_0 > + \sum_{k=1}^{n} < \Phi_k, \widetilde{S}_k > - < \Phi_{k-1}, \widetilde{S}_{k-1} >\end{aligned}$$

$$= < \Phi_0, \widetilde{S}_0 > + \sum_{k=1}^{n} < \Phi_k, \widetilde{S}_k > - < \Phi_k, \widetilde{S}_{k-1} >$$

$$= \widetilde{V}_0(\Phi) + \sum_{k=1}^{n} < \Phi_k, (\widetilde{S}_k - \widetilde{S}_{k-1}) > .$$

Conversely, if for any n, we have

$$\widetilde{V}_n(\Phi) = \widetilde{V}_0(\Phi) + \sum_{k=1}^{n} < \Phi_k, (\widetilde{S}_k - \widetilde{S}_{k-1}) >$$

$$= \widetilde{V}_0(\Phi) + \sum_{k=1}^{n-1} < \Phi_k, (\widetilde{S}_k - \widetilde{S}_{k-1}) > + < \Phi_n, (\widetilde{S}_n - \widetilde{S}_{n-1}) >$$

$$= \widetilde{V}_{n-1}(\Phi) + < \Phi_n, (\widetilde{S}_n - \widetilde{S}_{n-1}) >,$$

then, we also have

$$< \Phi_n, S_n > = < \Phi_{n-1}, S_{n-1} > + < \Phi_n, (\widetilde{S}_n - \widetilde{S}_{n-1}) >,$$

hence

$$< \Phi_{n-1}, S_{n-1} > = < \Phi_n, \widetilde{S}_{n-1} >,$$

which is indeed the self-financing relation. □

We have seen two important characteristics of investment strategies: the fact that it is predictable and self-financing. We now introduce a final constraint: the wealth must be non-negative at each instant.

DEFINITION 5.5.– *An investment strategy Φ is said to be* ***admissible*** *if it is*

– *predictable;*

– *self-financed;*

– *such that $V_n(\Phi) \geq 0$ for any $0 \leq n \leq N$.*

We may, however, have a negative number of shares for some assets (which corresponds to a debt) on the condition that the total wealth is always positive or zero (thus, it is possible to reimburse it at any instant).

5.3. Arbitrage

We will now establish a link between *viable* financial markets and martingales. The term *viable* is used here to signify the impossibility of a definite increase in wealth with an initial investment of zero. Let us formalize this.

DEFINITION 5.6.– *An investment strategy is said to be an* ***arbitrage strategy*** *or* ***arbitrage opportunity*** *if it is an admissible strategy, with an initial value of zero and a final, non-zero value:*

– $V_0(\Phi) = 0$;

– $V_n(\Phi) \geq 0$ *for any* $0 \leq n \leq N$;

– *there exist* n *and* $\omega \in \Omega$ *such that* $V_n(\omega) > 0$.

Let us note that the last condition would simply translate to $\mathbb{P}(V_n > 0) > 0$ if the universe Ω was not finite.

DEFINITION 5.7.– *A financial market is said to be* ***viable*** *if there are no arbitrage opportunities.*

We will now see that the absence of an arbitrage opportunity translates into the fact that discounted risky assets are martingales, except for a suitably changed probability.

DEFINITION 5.8.– *A probability* $\mathbb{P}^*$ *on* $(\Omega, \mathcal{F})$ *is a* ***risk-neutral probability*** *if, under this probability, all discounted risky assets are martingales.*

THEOREM 5.1.– *A financial market is viable if and only if there exists a risk-neutral probability* $\mathbb{P}^*$ *equivalent*[1] *to* $\mathbb{P}$.

PROOF.– It is assumed that $\mathbb{P}^*$ exists and that there exists an admissible strategy Φ such that $V_0(\Phi) = 0$. Under $\mathbb{P}^*$, $(\widetilde{S}_n)$ is a martingale and since Φ is self-financed, we have

$$\widetilde{V}_n(\Phi) = \widetilde{V}_0(\Phi) + \sum_{k=1}^{n} < \Phi_k, (\widetilde{S}_k - \widetilde{S}_{k-1}) > .$$

It can thus be seen that $(V_n(\Phi))$ is a martingale transform, since Φ is predictable; therefore, it is a martingale under $\mathbb{P}^*$. In particular, we have

$$\mathbb{E}[\widetilde{V}_n(\Phi)] = \mathbb{E}[\widetilde{V}_0(\Phi)] = 0$$

for any n. In addition, since $\widetilde{V}_n(\Phi) \geq 0$ for any n, because Φ is admissible, we necessarily have that $\widetilde{V}_n(\Phi(\omega) = 0$ for all $\omega \in \Omega$. Thus, no arbitrage opportunity exists.

1 On a finite set Ω, we say that $\mathbb{P}$ and $\mathbb{P}^*$ are equivalent if for any $\omega \in \Omega$, $\mathbb{P}^*(\omega) > 0 \Leftrightarrow \mathbb{P}(\omega) > 0$. More generally, on any probability space $(\Omega, \mathcal{F})$, $\mathbb{P}$ and $\mathbb{P}^*$ are equivalent if for any $A \in \mathcal{F}$, $\mathbb{P}^*(A) = 0 \Leftrightarrow \mathbb{P}(A) = 0$; in other words, the two probabilities yield the same zero-measure sets.

The converse is based on the hyperplane separation theorem and may be found in [LAM 97, Theorem 2.6], among others. As this does not use the probabilistic tools introduced in this book, and does not shed any special light on these tools or on financial markets, we will not include this proof here. □

5.4. The Cox, Ross and Rubinstein model

We will now illustrate the different concepts introduced above using a specific case of financial market. This is a discretized version of the Black and Scholes model.

The market is considered to be made up of a risk-free asset $S_n^0 = (1+r)^n$ and a single risky asset S^1 with the dynamic

$$S_0^1 = 1, \qquad S_{n+1}^1 = S_n^1 T_{n+1},\ n \geq 0,$$

where $(T_n)_{1\leq n\leq N}$ is a sequence of random variable taking only two values $1+d$ and $1+u$ with $-1 < d < u$. In other words, d and u are interpreted as interest rates. We thus write

$$\Omega = \{1+d, 1+u\}^N,\ \mathcal{F}_0 = \{\emptyset, \Omega\},\ \mathcal{F}_n = \sigma(T_1, \ldots, T_n),\ 1 \leq n \leq N.$$

In particular, $\mathcal{F}_N = \sigma(T_1, \ldots, T_N) = \mathcal{P}(\Omega)$ is the set of subsets of Ω.

We will now characterize viable markets in this model. In order to do this, we start by studying risk-neutral probabilities.

PROPOSITION 5.2.– *The discounted prices $(\widetilde{S}_n^1)_{0\leq n\leq N}$ are a martingale under a probability $\mathbb{P}^*$ if and only if, for any $0 \leq n \leq N-1$, we have*

$$\mathbb{E}^*[T_{n+1}|\mathcal{F}_n] = 1+r,$$

with $\mathbb{E}^*$ denoting the expectation for the probability $\mathbb{P}^*$.

PROOF.– Let us proceed through double implication.

Let us assume that the realized price is a martingale. For any $0 \leq n \leq N-1$, we then have

$$\begin{aligned}
&\mathbb{E}^*[\widetilde{S}_{n+1}^1|\mathcal{F}_n] = \widetilde{S}_n^1 \\
\Longrightarrow\ &\mathbb{E}^*[\frac{S_{n+1}^1}{(1+r)^{n+1}}|\mathcal{F}_n] = \frac{S_n^1}{(1+r)^n} \\
\Longrightarrow\ &\mathbb{E}^*[S_n^1 T_{n+1}|\mathcal{F}_n] = S_n^1(1+r)
\end{aligned}$$

$$\Longrightarrow S_n^1\mathbb{E}^*[T_{n+1}|\mathcal{F}_n] = S_n^1(1+r)$$
$$\Longrightarrow \mathbb{E}^*[T_{n+1}|\mathcal{F}_n] = (1+r),$$

since S_n^1 is the $\mathcal{F}_n$-measurable.

Conversely, if for any $0 \leq n \leq N-1$, we have $\mathbb{E}^*[T_{n+1}|\mathcal{F}_n] = 1+r$; therefore

$$\mathbb{E}^*[T_{n+1}|\mathcal{F}_n] = (1+r)$$
$$\Longrightarrow S_n^1\mathbb{E}^*[T_{n+1}|\mathcal{F}_n] = S_n^1(1+r)$$
$$\Longrightarrow \mathbb{E}^*[\frac{S_{n+1}^1}{(1+r)^{n+1}}|\mathcal{F}_n] = \frac{S_n^1}{(1+r)^n}$$
$$\Longrightarrow \mathbb{E}^*[\widetilde{S}_{n+1}^1|\mathcal{F}_n] = \widetilde{S}_n^1,$$

therefore $(\widetilde{S}_n^1)$ is indeed a martingale, as the measurability and integrability conditions are satisfied. □

We can now completely characterize viable markets in this model.

THEOREM 5.2.– *In the Cox, Ross and Rubinstein model, the market is viable if and only if the coefficients satisfy $d < r < u$.*

PROOF.– Let us assume that the market is viable. There then exists a risk-neutral probability $\mathbb{P}^*$ for which $(\widetilde{S}_n^1)$ is a martingale. According to the above proposition, we then have

$$1+r = \mathbb{E}^*[T_{n+1}|\mathcal{F}_n]$$
$$\Longrightarrow 1+r = \mathbb{E}^*[T_{n+1}] = (1-p_{n+1}^*)(1+d) + p_{n+1}^*(1+u)$$
$$\Longrightarrow r = \mathbb{E}^*[T_{n+1}] = (1-p_{n+1}^*)d + p_{n+1}^*u,$$

where we have written $p_{n+1}^* = \mathbb{P}^*(T_{n+1} = 1+u)$. Furthermore, $\mathbb{P}^*(\omega) > 0$ for all $\omega \in \Omega$ (since it is equivalent to $\mathbb{P}$, which verifies this same property); therefore, we have $0 < p_{n+1}^* < 1$, which does imply $d < r < u$.

Let us now assume that the market is not viable. There thus exists an arbitrage opportunity Φ with either $V_0(\Phi) = 0 = S_0^0 - S_0^1$, or $V_0(\Phi) = 0 = -S_0^0 + S_0^1$ (up to a multiplicative constant). In the first case, we have

$$V_0(\Phi) = 0 = S_0^0 - S_0^1,$$
$$V_1(\Phi)(\omega) = \begin{cases} 1+r-(1+u) = r-u \geq 0 \\ 1+r-(1+d) = r-d, \end{cases}$$

as per the values of ω. We then have $r - d > r - u \geq 0$; therefore, $r \geq u$. The other possibility is

$$V_0(\Phi) = 0 = -S_0^0 + S_0^1,$$

$$V_1(\Phi)(\omega) = \begin{cases} -(1+r) + 1 + u = u - r \\ -(1+r) + 1 + d = d - r \geq 0, \end{cases}$$

as per the values of ω. We then have $u - r > d - r \geq 0$; therefore, $r \leq d$. □

This result simply states that to have a viable market, it is necessary and sufficient for one of the increment values of the risky asset $(1+d)$ to be unfavorable with respect to the risk-free interest rate $(1+r)$, and that the other $(1+u)$ to be more favorable than the risk-free case. It can be easily understood that if both values u and d are smaller than the risk-free rate, nobody would invest in the risky asset. On the contrary, if both values u and d are greater than the risk-free rate, it is always seen as advantageous to invest all one's wealth in the risky asset.

5.5. Exercises

EXERCISE 5.1.– Consider the following market model on $\Omega = (\omega_1, \omega_2)$:

	$n = 0 : (S_0^0, S_0^1)$	$n = 1 : (S_1^0, S_1^0)$
ω_1	$(1, 2)$	$(1.05, 2.5)$
ω_2	$(1, 2)$	$(1.05, 5)$

1) Construct the event tree representing the evolution of this financial market.

2) Explicitize $\mathcal{F}_0$ and $\mathcal{F}_1$, the σ-algebras representing the information available at each instant.

3) Calculate the interest rate of the risk-free asset.

4) Calculate the acquisition cost of the initial portfolio of an investor who is short selling four shares in a risk-free asset (i.e. borrowing from the risk-free asset) to buy two shares in risky assets.

5) Calculate the possible values of the portfolio after the evolution of the asset values, and after reimbursing the borrowed risk-free assets.

6) Is this market viable?

EXERCISE 5.2.– Consider the following market model on $\Omega = \{\omega_1, \omega_2\}$

	$n = 0 : (S_0^0, S_0^1)$	$n = 1 : (S_1^0, S_1^1)$	$n = 2 : (S_2^0, S_2^1)$
ω_1	$(1, 1)$	$(1.1, 1.2)$	$(1.2, 1.5)$
ω_2	$(1, 1)$	$(1.1, 1.5)$	$(1.2, 1.2)$

1) Consider the strategy $\Phi_1(\omega_1) = \Phi_1(\omega_2) = (-1, 2)$, $\Phi_2(\omega_1) = \Phi_2(\omega_2) = (-1, 1)$. Calculate the wealth at every instant, corresponding to this strategy. Is this strategy admissible?

2) Consider the strategy $\Phi_1(\omega_1) = \Phi_1(\omega_2) = (-1, 2)$, $\Phi_2(\omega_1) = \Phi_2(\omega_2) = (-2, 4)$. Calculate the wealth at every instant corresponding to this strategy. Is this strategy admissible?

3) Consider the strategy $\Phi_1(\omega_1) = \Phi_1(\omega_2) = (-1, 1)$, $\Phi_2(\omega_1) = (0.08, 0.01)$, $\Phi_2(\omega_2) = (-16, 12)$. Calculate the wealth at every instant, corresponding to this strategy. Is this strategy admissible?

4) Consider the strategy $\Phi_1(\omega_1) = \Phi_1(\omega_2) = (-2, 2)$, $\Phi_2(\omega_1) = \Phi_2(\omega_2) = (-2, 2)$. Calculate the wealth at every instant corresponding to this strategy. Is this strategy admissible?

5) Is this market viable?

EXERCISE 5.3.– Consider the following market model on $\Omega = (\omega_1, \omega_2, \omega_3)$:

	$n = 0 : (S_0^0, S_0^1, S_0^2)$	$n = 1 : (S_1^0, S_1^1, S_1^2)$
ω_1	$(1, 1, 1)$	$(1.1, 1.65, 0.88)$
ω_2	$(1, 1, 1)$	$(1.1, 1.65, 0.88)$
ω_3	$(1, 1, 1)$	$(1.1, 0.55, 1.32)$

1) Construct the event tree representing the evolution of this financial market.

2) Explicitize $\mathcal{F}_0$ and $\mathcal{F}_1$, the σ-algebras representing the information available at each instant.

3) Show that the probability $\mathbb{P}_1(\omega_1) = 1/4$, $\mathbb{P}_1(\omega_2) = 1/4$, $\mathbb{P}_1(\omega_3) = 1/2$ is a risk-neutral probability.

4) Show that the probability $\mathbb{P}_2(\omega_1) = 2/10$, $\mathbb{P}_2(\omega_2) = 3/10$, $\mathbb{P}_2(\omega_3) = 1/2$ is a risk-neutral probability.

5) Is this market viable?

EXERCISE 5.4.– Construct an arbitrage opportunity in a market following the Cox, Ross and Rubinstein model with $d \geq r$.

EXERCISE 5.5.– Construct an arbitrage opportunity in a market following the Cox, Ross and Rubinstein model with $u \leq r$.

5.6. Practical work

We work with a Cox, Ross and Rubinstein model, also assuming that the sequence (T_n) is i.i.d. We write

$$\mathbb{P}(T_1 = 1 + u) = 1 - \mathbb{P}(T_1 = 1 + d) = q.$$

5.6.1. *Simulation of trajectories*

Initially, we wish to simulate the trajectories of the risky asset S^1.

1) Write a function `CRR(N,u,d,q)` which samples a trajectory on $[0, \texttt{N}]$ of a risky asset, which evolves as per the Cox, Ross and Rubinstein model with the parameters `u`, `d` and `q`. You can use the function R `sample` as well as the command `cumprod`.

2) Graphically represent some trajectories for $r = 0$, $u = 10\%$, $d = -5\%$, $q = 1/3$ and $N = 50$.

5.6.2. *Portfolio optimization*

We now study a portfolio optimization problem in the Cox, Ross and Rubinstein model.

Let V_0 be the wealth of an investor at the time 0. The investor can invest their money either in a risky asset or in a risk-free asset, following an admissible strategy. We use ϕ_n^0 and ϕ_n^1 to denote the number of shares in the risk-free asset and the number of shares in the risky asset, respectively, held between the time $n-1$ and n. Let π_n be the *proportion* of the wealth invested in the risky asset between the instants $n-1$ and n, that is,

$$\pi_n = \frac{\phi_n^1 S_{n-1}^1}{V_{n-1}}.$$

1) Express ϕ_n^0 and ϕ_n^1 as the functions of π_n, S_{n-1}^0, S_{n-1}^1 and V_{n-1} for any n.

2) Derive from this that for any n, the wealth at the time n, after the evolution of the prices and before the redistribution of the portfolio has the value:

$$V_n = \big(\pi_n T_n + (1 - \pi_n)(1 + r)\big) V_{n-1}.$$

3) On the same graph and for the **same random sampled trajectory**, represent the evolution of the risk-free asset and the evolution of the wealth for the following two strategies:

a) The *proportion* of the wealth invested in the risky asset is fixed over time, at $1/4$.

b) The *proportion* of the wealth invested in the risky asset only takes the values 0 and 1. It takes the value 1 when the price of a risky asset strictly exceeds that of the risk-free asset, and takes the value 0 when the risky asset is small than or equal to the risk-free asset, *while remaining predictable.*

We will take the following parameters: initial wealth $V_0 = 1$, the risky asset evolves as per the Cox, Ross and Rubinstein model with parameters $d = -2\%$, $u = 10\%$ and $q = 0.52$, interest rate $r = 4\%$ and duration of investment: $N = 100$ periods.

Now consider that the investor's utility function is logarithmic and we wish to find the strategy $\pi^* = (\pi_n^*)_{1 \leq n \leq N}$ which maximizes the expectation of the utility of the wealth at maturity N :

$$\sup_{\pi \text{ admissible}} \mathbb{E}[\log V_N(\pi)].$$

We will accept that the optimal strategy is constant over time and maximize the expression

$$q \log(\pi^*(u-r)+1+r)+(1-q)\log(\pi^*(d-r)+1+r).$$

We wish to compare the performances of strategies 1 and 2, given above, and that of the optimal strategy. We will use the same parameters as for the previous question.

4) Write a function `optimal(u,d,r,q)`, which calculates the value of π^* the optimal proportion to invest in the risky asset. We can restrict ourselves to five decimals. What do we find for our parameters?

5) On the same graph, trace a trajectory of the *logarithm* of the wealth at each instant for the optimal strategy, and for strategies 1 and 2. Does the optimal strategy always give the same result? Why?

6) On the same graph, trace the expectation of the *logarithm* of the wealth at each instant, for each of the three strategies. We will calculate the expectation using the Monte Carlo method. Which is the best strategy? Discuss.

5.6.3. *Portfolio optimization with withdrawal*

In this section, the investor is allowed to withdraw a *proportion* c_n of their wealth at the instant n, *after* updating the asset prices, but *before* the redistribution of their portfolio for the next investment period. Therefore, they only reinvest the non-withdrawn part. The corresponding investment-withdrawal strategy (π_n, c_n) is no longer self-financed, but it must remain predictable and the wealth after the withdrawal must be positive or zero at each instant. Therefore, it can thus be shown that the new wealth at the time n after the evolution of the prices and when the value of the withdrawal is

$$V_n(\pi, c) = \prod_{i=1}^{n} (1 - c_i)\big(\pi_i T_i + (1 - \pi_i)(1 + r)\big),$$

such that the value of the wealth withdrawn at the instant n is $R_n(\pi, c)$ with

$$R_n(\pi, c) = \begin{cases} \frac{c_n V_n(\pi,c)}{1-c_n} & \text{if } c_n \neq 1, \\ V_{n-1}(\pi, c)\big(\pi_n T_n + (1 - \pi_n)(1 + r)\big) & \text{if not.} \end{cases}$$

1) Graphically represent the evolution of the wealth for the investment strategy with the following withdrawal policy:

a) The proportion of the wealth invested in the risky asset is fixed over time, at $1/4$,

b) The proportion of the wealth withdrawn at each instant is equal to 1% over the interval $[1, 80]$, 5% over the interval $]80, 90]$, 10% over the interval $]90, 95]$, 25% over the interval $]95, 100[$ and upon maturity, all the remaining wealth is withdrawn.

We will take the following parameters: the initial wealth $V_0 = 1$, the risky asset evolves as per the Cox, Ross and Rubinstein model with parameters $d = -2\%$, $u = 10\%$ and $q = 0.52$, interest rate $r = 4\%$, duration of investment: $N = 100$ periods.

We now consider, once again, that the investor's utility function is logarithmic and we want to find the investment strategy with the withdrawal (π_n, c_n) that maximizes the expectation of the cumulative sum of the withdrawal utility up to the date of maturity N:

$$\max_{(\pi_n, c_n)} \mathbb{E}\Big[\sum_{n=1}^{N} \log\big(R_n(\pi, c)\big)\Big].$$

We will admit that the investment strategy with optimal withdrawal is given by

- $\pi_n = \pi^*$ for any $1 \leq n \leq N$,

- $c_n = \frac{1}{N+1-n}$ for any $1 \leq n \leq N$,

with the same π^* as in the earlier practical exercise.

We wish to compare the performances of the strategy in the above exercise and that of the optimal strategy. We will use the same parameters as for the previous exercise.

2) Write a function `strategies_conso` that calculates, for the *same* random sampled trajectory, the wealth at every instant for the optimal strategy, and the above strategy.

3) On the same graph, trace a trajectory for the wealth at each instant for each of these two strategies.

4) Calculate at each instant n, for the same sampled trajectory, and for each of the two strategies, the value of the cumulative sum of the logarithm of the withdrawal up to the instant n. On the same graphic, trace the two trajectories of the cumulative sum for the square roots of withdrawals between 0 and N. Is the optimal strategy the same for all samples? Why?

5) On the same graph, trace the *expectation* of the cumulative sum of the logarithm of the withdrawal up to the date of maturity N for the two strategies. We will calculate the expectation using the Monte Carlo method by taking the average of 1,000 iterations of each trajectory. Which is the best strategy? Discuss.

6

European Options

In this chapter, we will study in detail a specific form of a financial contract or conditional asset called a *European option*. Section 6.1 defines a European option and its characteristics. Section 6.2 introduces the concept of a complete financial market and its characterization in terms of a martingale. Section 6.3 resolves the problem of the valuation and hedging of European options in a general framework. Section 6.4 applies these results in a particular case of the Cox, Ross and Rubinstein model. Finally, the chapter ends with exercises in section 6.5 and practical work in section 6.6.

We still assume that Ω has a finite cardinal and that $\mathbb{P}(\omega) > 0$ for any $\omega \in \Omega$. Let N be a time horizon and let $(\mathcal{F}_n)_{0\leq n\leq N}$ be a filtration such that $\mathcal{F}_0 = \{\emptyset, \Omega\}$ and $\mathcal{F}_N = \mathcal{F}$.

6.1. Definition

We start by writing the definition of a European option and giving the two chief examples: the put options and call options.

DEFINITION 6.1.– *A European option with maturity N is a random $X_N \geq 0$ $\mathcal{F}_N$-measurable variable representing the profit that makes it possible to exercise the option at the instant N.*

The fact that an option X_N is the $\mathcal{F}_N$-measurable means that its value is not known (*a priori*) before its date of maturity N. The decisions made by investors with respect to European options are thus made under uncertainty.

EXAMPLE 6.1.– *A European call option gives its holder the right, but not the obligation, to buy, at the date of maturity N, a share of a the underlying risk-asset $S = (S_n)_{0\leq n\leq N}$ at a strike price K, determined in advance.*

– *If* $S_N < K$, *the option is not exercised and the payoff is zero;*

– *if* $S_N \geq K$, *the option is exercised and the payoff is* $S_N - K$.

The associated random variable is therefore $X_N = (S_N - K)_+ = \max\{S_N - K, 0\}$. *Such a contract makes it possible to hedge against inflation of the asset that we wish to buy.* ♢

EXAMPLE 6.2.– *A European put option gives its holder the right, but not the obligation, to sell a share of the underlying risky asset* $S = (S_n)_{0 \leq n \leq N}$, *upon the date of maturity* N, *at a strike price* K, *determined in advance.*

– *If* $S_N > K$, *the option is not exercised and the payoff is zero;*

– *if* $S_N \leq K$, *the option is exercised and the payoff is* $K - S_N$.

The random variable associated with the payoff is therefore $X_N = (K - S_N)_+$. *Such a contract makes it possible to hedge against the fall in price of the asset that we wish to sell.* ♢

There are also other types of options, such as:

– American options, which may be exercised at any point up to the date of maturity, see Chapter 7.

– Asian options, which depend on the mean value of the asset over a time window.

– Swap options, which have two underlying assets that may (or may not) be exchanged on certain dates,

and many others, called *exotic options* (see, for example, [HUL 18]).

6.2. Complete markets

We now wish to see whether it is possible to construct admissible strategies that allow us to obtain a European option X_N at the date of maturity N.

DEFINITION 6.2.– *A European option* X_N *is said to be* ***simulable*** *if there exists an admissible strategy whose value at the date of maturity* N *is exactly* X_N.

DEFINITION 6.3.– *A market is said to be* ***complete*** *if any European option is simulable.*

We will now see that it is possible to characterize complete markets using martingales.

THEOREM 6.1.– *A market is viable and complete if and only if there exists a* ***unique*** *risk-neutral probability* $\mathbb{P}^*$ *equivalent to* $\mathbb{P}$.

PROOF.– The characterization of a viable market through the existence of viable, risk-neutral probabilities has already been seen in Theorem 5.2.

Let us assume that the market is viable and complete and that there exist two risk-neutral probabilities $\mathbb{P}_1$ and $\mathbb{P}_2$. Let X_N be a European option. Since the market is complete, there exists an admissible strategy Φ such that $V_N(\Phi) = X_N$. Furthermore, under a risk-neutral probability, the discounted value of the portfolio $(\widetilde{V}_n(\Phi))$ is a martingale. In particular, we have

$$\mathbb{E}_1[V_N(\Phi)] = \frac{\mathbb{E}_1[X_N]}{S_N^0} = \mathbb{E}_1[\widetilde{V}_0(\Phi)] = V_0(\Phi)$$

$$\mathbb{E}_2[V_N(\Phi)] = \frac{\mathbb{E}_2[X_N]}{S_N^0} = \mathbb{E}_2[\widetilde{V}_0(\Phi)] = V_0(\Phi)$$

Thus, for any random X_N $\mathcal{F}_N = \mathcal{F}$-measurable variable, we have $\mathbb{E}_1[X_N] = \mathbb{E}_2[X_N]$. Taking $X_N = \mathbb{1}_\omega$, for example, we obtain $\mathbb{P}_1(\omega) = \mathbb{P}_2(\omega)$ for any $\omega \in \Omega$; therefore, $\mathbb{P}_1 = \mathbb{P}_2$. Hence, we have unicity.

The converse is based on the orthogonality properties of the sub-spaces of random variables and sheds no particular light on the concepts of martingales or financial markets. We have therefore omitted it here and the reader who wishes to learn more about them may wish to consult, among others, [LAM 97, Theorem 3.4] for more details. □

6.3. Valuation and hedging

In all that follows, it is assumed that the market is viable and complete, $\mathbb{P}^*$ denotes the unique risk-neutral probability and $\mathbb{E}^*$ denotes the expectation under $\mathbb{P}^*$. It is therefore known that there exist admissible strategies that make it possible to obtain the European option X_N at the date of maturity. We now wish to find the initial wealth needed for these investment strategies.

DEFINITION 6.4.– Φ *is said to be a* ***hedging strategy*** *for the option* X_N*, if it is an admissible strategy such that* $V_N(\Phi) = X_N$.

DEFINITION 6.5.– *The valuation of an option* X_N *means finding the value* $V_0(\Phi)$ *of the portfolio that must be held at the initial instant to obtain the option* X_N *at the date of maturity* N.

When the market is complete it is known that hedging strategies exist, but there is no guarantee of unicity. The value of an option, *a priori*, could depend on the choice of the hedging strategy. However, we will see that this is not actually the case.

PROPOSITION 6.1.– *The value of an option does not depend on the choice of hedging strategy. More precisely, the value of an option X_N with the maturity date N, at the instant n for $0 \leq n \leq N$ and for any hedging strategy Φ is*

$$V_n(\Phi) = (1+r)^{n-N}\mathbb{E}^*[X_N|\mathcal{F}_n],$$

where r is the interest rate of the risk-free asset.

PROOF.– Let Φ be a hedging strategy for X_N. Since we have a risk-neutral probability, the discounted value of the portfolio $(\widetilde{V}_n(\Phi))_{0\leq n\leq N}$ is a martingale. Therefore, we have

$$\widetilde{V}_n(\Phi) = \mathbb{E}^*[\widetilde{V}_N(\Phi)|\mathcal{F}_n] = \frac{\mathbb{E}^*[X_N|\mathcal{F}_n]}{S_N^0} = \frac{\mathbb{E}^*[X_N|\mathcal{F}_n]}{(1+r)^N}.$$

Since we also have $V_n(\Phi) = \widetilde{V}_n(\Phi)S_n^0 = (1+r)^n\widetilde{V}_n(\Phi)$, we obtain the stated result. □

REMARK 6.1.– *It may be seen in particular that according to the above formula, the value of an option depends only on the risk-neutral probability $\mathbb{P}^*$ and not the historical probability $\mathbb{P}$.* ★

EXAMPLE 6.3.– *For a call option with the strike price K, date of maturity N and with the underlying asset $S = (S_n)_{0\leq n\leq N}$, we obtain*

$$V_n(\Phi) = C_n = (1+r)^{n-N}\mathbb{E}^*[(S_N - K)_+|\mathcal{F}_n],$$

and for a put option with the same characteristics, we have

$$V_n(\Phi) = P_n = (1+r)^{n-N}\mathbb{E}^*[(K - S_N)_+|\mathcal{F}_n].$$

If the underlying asset dynamic is not specified, it is not possible to go further on generalities. ◇

6.4. Cox, Ross and Rubinstein model

In the specific case of the Cox, Ross and Rubinstein model, it is possible to go further in the study of European options and obtain a closed-form formula for the value of the option at each instant, as well as an explicit construction of a hedging strategy.

6.4.1. *Completeness*

In the following, we work with a viable Cox, Ross and Rubinstein market model: $d < r < u$. We will start by finding the conditions under which the market is also complete.

THEOREM 6.2.– *Let* $p = \frac{r-d}{u-d}$. *Then, the discounted risky asset* $(\widetilde{S}^1_n)_{0 \le n \le N}$ *is a martingale under* $\mathbb{P}^*$ *if and only if the multiplicative increments* $T_1, \ldots, T_N$

– *are independent under* $\mathbb{P}^*$ *and*

– *are identically distributed under* $\mathbb{P}^*$, *given by* $\mathbb{P}^*(T_n = 1 + u) = p$.

PROOF.– Let us assume that T_n are i.i.d. in the distribution given by p. We then have by independence, since $\mathcal{F}_n = \sigma(T_1, \ldots, T_n)$

$$\begin{aligned}\mathbb{E}^*[T_{n+1}|\mathcal{F}_n] &= \mathbb{E}^*[T_{n+1}] \\ &= (1-p)(1+d) + p(1+u) \\ &= \frac{u-r}{u-d}(1+d) + \frac{r-d}{u-d}(1+u) \\ &= 1 + r.\end{aligned}$$

Therefore, $(\widetilde{S}^1_n)$ is a martingale under $\mathbb{P}^*$ using Proposition 5.2.

Conversely, if $(\widetilde{S}^1_n)$ is a martingale under $\mathbb{P}^*$, since T_{n+1} can only take the two values $1 + d$ and $1 + u$, we have

$$\begin{aligned}&\mathbb{E}^*[T_{n+1}|\mathcal{F}_n] \\ &= 1 + r \\ &= (1+d)\mathbb{E}^*[\mathbb{1}_{(T_{n+1}=1+d)}\,|\mathcal{F}_n] + (1+u)\mathbb{E}^*[\mathbb{1}_{(T_{n+1}=1+u)}\,|\mathcal{F}_n] \\ &= (1+d)\mathbb{E}^*[1 - \mathbb{1}_{(T_{n+1}=1+u)}\,|\mathcal{F}_n] + (1+u)\mathbb{E}^*[\mathbb{1}_{(T_{n+1}=1+u)}\,|\mathcal{F}_n] \\ &= 1 + d + (u-d)\mathbb{P}^*(T_{n+1} = 1 + u|\mathcal{F}_n).\end{aligned}$$

Moving to the expectations of both sides, we have

$$1 + r = 1 + d + (u-d)\mathbb{P}^*(T_{n+1} = 1 + u),$$

therefore, $\mathbb{P}^*(T_{n+1} = 1 + u) = p$. We have thus identified the distribution of the T_n.

For independence, we proceed by induction on n.

For $n = 2$, for any $(x, y) \in \{1 + d, 1 + u\}^2$, we write $p(x) = p$ if $x = 1 + u$ and $p(x) = 1 - p$ if $x = 1 + d$. Using the properties of conditional expectation, it follows

$$\begin{aligned}\mathbb{P}^*(T_1 = x, T_2 = y) &= \mathbb{E}^*[\mathbb{1}_{(T_1=x)}\mathbb{1}_{(T_2=y)}] \\ &= \mathbb{E}^*\left[\mathbb{E}^*[\mathbb{1}_{(T_1=x)}\mathbb{1}_{(T_2=y)}|\mathcal{F}_1]\right] \\ &= \mathbb{E}^*\left[\mathbb{1}_{(T_1=x)}\mathbb{E}^*[\mathbb{1}_{(T_2=y)}|\mathcal{F}_1]\right]\end{aligned}$$

$$
\begin{aligned}
&= \mathbb{E}^*\left[\mathbb{1}_{(T_1=x)}p(y)\right] \\
&= p(y)\mathbb{E}^*\left[\mathbb{1}_{(T_1=x)}\right] \\
&= p(y)p(x) \\
&= \mathbb{P}^*(T_1=x)\mathbb{P}^*(T_2=y).
\end{aligned}
$$

Thus, T_1 and T_2 are independent.

Let us assume that for a fixed n, we have

$$
\mathbb{P}^*(T_1=x_1,\dots T_n=x_n)=\prod_{k=1}^{n}\mathbb{P}^*(T_k=x_k).
$$

Thus, for step $n+1$, we have

$$
\begin{aligned}
&\mathbb{P}^*(T_1=x_1,\dots T_n=x_n,T_{n+1}=x_{n+1}) \\
&= \mathbb{E}^*\left[\mathbb{1}_{(T_1=x_1,\dots T_n=x_n)}\mathbb{1}_{(T_{n+1}=x_{n+1})}\right] \\
&= \mathbb{E}^*\left[\mathbb{E}^*\left[\mathbb{1}_{(T_1=x_1,\dots T_n=x_n)}\mathbb{1}_{(T_{n+1}=x_{n+1})}|\mathcal{F}_n\right]\right] \\
&= \mathbb{E}^*\left[\mathbb{1}_{(T_1=x_1,\dots T_n=x_n)}\mathbb{E}^*\left[\mathbb{1}_{(T_{n+1}=x_{n+1})}|\mathcal{F}_n\right]\right] \\
&= \mathbb{E}^*\left[\mathbb{1}_{(T_1=x_1,\dots T_n=x_n)}p(x_{n+1})\right] \\
&= p(x_{n+1})\mathbb{E}^*\left[\mathbb{1}_{(T_1=x_1,\dots T_n=x_n)}\right] \\
&= p(x_{n+1})\prod_{k=1}^{n}\mathbb{P}^*(T_k=x_k) \\
&= \prod_{k=1}^{n+1}\mathbb{P}^*(T_k=x_k).
\end{aligned}
$$

Hence, we have independence. □

The following result can immediately be derived from this.

COROLLARY 6.1.– *The Cox, Ross and Rubinstein market is viable and complete if and only if $d < r < u$ and the unique risk-neutral probability is characterized by p.*

PROOF.– From Theorem 5.2, it is known that the market is viable if and only if the condition $d < r < u$ is satisfied. According to what we have seen before, if the market is viable, then there is a unique risk-neutral probability characterized by p. We conclude by using Theorem 6.1. □

6.4.2. *Value of European options*

We now seek to find the value of call options and put options in the Cox, Ross and Rubinstein model. We use C_n and P_n to denote the value of a European call option and a European put option, respectively, at the time n in the Cox, Ross and Rubinstein model, with the maturity date N, a strike price K and the underlying asset $(S_n^1)_{0\le n\le N}$.

PROPOSITION 6.2.– *The value of the call option is* $C_n = c(n, S_n^1)$, *where* c *is the deterministic function defined by*

$$c(n,x) = (1+r)^{n-N} \sum_{j=0}^{N-n} C_{N-n}^j p^j (1-p)^{N-n-j} \left(x(1+u)^j(1+d)^{N-n-j} - K\right)_+.$$

PROOF.– Using the general formula for options in Proposition 6.1, it is known that we have

$$\begin{aligned} C_n &= (1+r)^{n-N}\mathbb{E}^*[(S_N^1 - K)_+|\mathcal{F}_n] \\ &= (1+r)^{n-N}\mathbb{E}^*[(S_n^1 T_{n+1}\cdots T_N - K)_+|\mathcal{F}_n]. \end{aligned}$$

Furthermore, under $\mathbb{P}^*$, the T_n are independent; therefore, $T_{n+1}\cdots T_N$ is independent of $\mathcal{F}_n$. In addition, S_n^1 is the $\mathcal{F}_N$-measurable variable. Using Proposition 2.13, we can thus write

$$C_n = (1+r)^{n-N} H(S_n^1),$$

with

$$H(x) = \mathbb{E}^*[(xT_{n+1}\cdots T_N - K)_+].$$

The product $T_{n+1}\cdots T_N$ takes the values $(1+u)^j(1+d)^{N-n-j}$ with the probability $C_{N-n}^j p^j(1-p)^{N-n-j}$ under $\mathbb{P}^*$. Therefore, we do indeed have

$$H(x) = \sum_{j=0}^{N-n} C_{N-n}^j p^j (1-p)^{N-n-j} \left(x(1+u)^j(1+d)^{N-n-j} - K\right)_+,$$

hence, the result. □

The above formula is completely explicit; however, it is potentially unwieldy for numerical calculations. For these calculations, we prefer using the following version, which gives a backward recurrence relation between the C_n.

COROLLARY 6.2.– *The value of the call option is* $C_n = c(n, S_n^1)$*, where* c *satisfies the recurrence relation*

$$C(N, x) = (x - K)_+$$
$$c(n, x) = (1 + r)^{-1}\big(c(n + 1, (1 + u)x)p + c(n + 1, (1 + d)x)(1 - p)\big).$$

PROOF.– As we have an explicit formula, it is enough to verify that the given expression coincides with the explicit formula. As per the previous formula, we have

$$c(N, x) = (1 + r)^{N-N}\sum_{j=0}^{N-N} C_{N-N}^{j} p^j (1 - p)^{N-N-j}\big(x(1 + u)^j (1 + d)^{N-N-j} - K\big)_+$$
$$= (x - K)_+$$

since the sum contains only a single term, $j = 0$, and by shifting the index of the first summation, we also obtain

$$(1 + r)^{N-n-1}\big(c(n + 1, (1 + u)x)p + c(n + 1, (1 + d)x)(1 - p)\big)$$
$$= \sum_{j=0}^{N-n-1} C_{N-n-1}^{j} p^{j+1}(1 - p)^{N-n-1-j}\big(x(1 + u)^{j+1}(1 + d)^{N-n-1-j} - K\big)_+$$
$$+ \sum_{j=0}^{N-n-1} C_{N-n-1}^{j} p^{j}(1 - p)^{N-n-1-j+1}\big(x(1 + u)^{j}(1 + d)^{N-n-1-j+1} - K\big)_+$$
$$= \sum_{j=1}^{N-n} C_{N-n-1}^{j+1} p^{j}(1 - p)^{N-n-j}\big(x(1 + u)^{j}(1 + d)^{N-n-j} - K\big)_+$$
$$+ \sum_{j=0}^{N-n-1} C_{N-n-1}^{j} p^{j}(1 - p)^{N-n-j}\big(x(1 + u)^{j}(1 + d)^{N-n-j} - K\big)_+$$
$$= \sum_{j=1}^{N-n-1}(C_{N-n-1}^{j+1} + C_{N-n-1}^{j})p^{j}(1 - p)^{N-n-j}\big(x(1 + u)^{j}(1 + d)^{N-n-j} - K\big)_+$$
$$+ p^{N-n}(x(1 + u)^{N-n} - K)_+ + (1 + r)^{n+N}\left((1 - p)^{N-n}(x(1 + d)^{N-n} - K)_+\right.$$
$$= \sum_{j=0}^{N-n} (C_{N-n-1}^{j+1} + C_{N-n-1}^{j})p^{j}(1 - p)^{N-n-j}\big(x(1 + u)^{j}(1 + d)^{N-n-j} - K\big)_+$$

$$= \sum_{j=0}^{N-n} C_{N-n}^{j} p^j (1-p)^{N-n-j} \left(x(1+u)^j(1+d)^{N-n-j} - K\right)_+$$

$$= (1+r)^{N-n} c(n,x),$$

hence the result. □

We can obtain the value of a put option using the same steps.

PROPOSITION 6.3.– *The value of the put option with maturity date N, strike price K and underlying asset S^1 is $P_n = v(n, S_n^1)$, where v is the deterministic function defined by*

$$v(n,x) = (1+r)^{n-N} \sum_{j=0}^{N-n} C_{N-n}^{j} p^j (1-p)^{N-n-j} \left(K - x(1+u)^j(1+d)^{N-n-j}\right)_+.$$

In addition, the function v satisfies the recurrence relation

$$v(N,x) = (K-x)_+$$

$$v(n,x) = (1+r)^{-1}\big(v(n+1,(1+u)x)p + v(n+1,(1+d)x)(1-p)\big).$$

These results may also be directly obtained from the value of a call option, using the *Call–Put parity relation*, which is used in Exercise 6.2 .

6.4.3. *Hedging European options*

We finally study the construction of an explicit hedging strategy.

PROPOSITION 6.4.– *It is possible to explicitly construct an admissible strategy that allows the hedging of a European call option in the Cox, Ross and Rubinstein model. The number of shares held between the instant n and the instant $n+1$ are*

$$\phi_{n+1}^0 = \Delta^0(n+1, S_n^1)$$

$$\phi_{n+1}^1 = \Delta^1(n+1, S_n^1)$$

for risk-free and risky assets, respectively, where Δ^0 and Δ^1 are the deterministic functions defined by

$$\Delta^0(n,x) = \frac{(1+u)c(n,x(1+d)) - (1+d)c(n,x(1+u))}{(1+r)^n(u-d)}$$

$$\Delta^1(n,x) = \frac{c(n,x(1+d)) - c(n,x(1+u))}{x(u-d)},$$

using the function c from Proposition 6.2.

PROOF.– Since the market is complete, there exists an admissible strategy Φ such that

$$\phi_n^0(1+r)^n + \phi_n^1 S_n^1 = C_n = c(n, S_n^1).$$

S_n^1 can be decomposed into the product $S_{n-1}^1 T_n$, and then multiplied by $\mathbb{1}_{(T_n=1+u)}$ and we take the conditional expectation, given $\mathcal{F}_{n-1}$, of both sides. We thus obtain

$$\begin{aligned}
&\phi_n^0(1+r)^n + \phi_n^1 S_{n-1}^1 T_n = c(n, S_{n-1}^1 T_n)\\
&\quad\Longrightarrow \mathbb{E}^*\left[\left(\phi_n^0(1+r)^n + \phi_n^1 S_{n-1}^1(1+u)\right)\mathbb{1}_{(T_n=1+u)}|\mathcal{F}_{n-1}\right]\\
&\quad\quad = \mathbb{E}^*\left[c(n, S_{n-1}^1(1+u))\mathbb{1}_{(T_n=1+u)}|\mathcal{F}_{n-1}\right]\\
&\quad\Longrightarrow \left(\phi_n^0(1+r)^n + \phi_n^1 S_{n-1}^1(1+u)\right)\mathbb{P}^*(T_n = 1+u)\\
&\quad\quad = c(n, S_{n-1}^1(1+u))\mathbb{P}^*(T_n = 1+u)\\
&\quad\Longrightarrow \phi_n^0(1+r)^n + \phi_n^1 S_{n-1}^1(1+u) = c(n, S_{n-1}^1(1+u)).
\end{aligned}$$

By following the same steps again, but this time multiplying by $\mathbb{1}_{(T_n=1+d)}$, we obtain

$$\phi_n^0(1+r)^n + \phi_n^1 S_{n-1}^1(1+d) = c(n, S_{n-1}^1(1+d)).$$

We thus have a system of two equations with two unknowns ϕ_n^0 and ϕ_n^1, whose solution yields the desired result. □

Once again, the same relation can be obtained for the hedging strategy of a put option by replacing the function c by the function v.

6.5. Exercises

EXERCISE 6.1.– Consider the following market model on $\Omega = \{\omega_1, \omega_2, \omega_3\}$

	$n = 0 : (S_0^0, S_0^1, S_0^2)$	$n = 1 : (S_1^0, S_1^1, S_1^2)$
ω_1	$(1, 1, 1)$	$(1.1, 0.66, 1.32)$
ω_2	$(1, 1, 1)$	$(1.1, 1.54, 0.88)$
ω_3	$(1, 1, 1)$	$(1.1, 1.54, 0.88)$

1) Calculate the interest rate for the risk-free asset and the discounted price of the risky assets.

2) Consider the option $X_1(\omega_1) = X_1(\omega_2) = 1$, $X_1(\omega_3) = 1.1$. Is it simulable?

3) Is this market complete?

4) Calculate $\mathbb{E}[\widetilde{S}_1^1 \mid \mathcal{F}_0]$ and $\mathbb{E}[\widetilde{S}_1^2 \mid \mathcal{F}_0]$ for the following two probabilities: $\mathcal{P}_1(\omega_2) = \mathcal{P}_1(\omega_3) = 0.25$, and $\mathcal{P}_2(\omega_2) = 0.2$, $\mathcal{P}_2(\omega_3) = 0.3$.

5) Is this market viable?

EXERCISE 6.2.– Demonstrate the Call–Put parity relation in a viable and complete market: for the call options and put options with the same maturity date N, with the same strike price K and the same underlying asset S, we have, for any $0 \leq n \leq N$,

$$C_n - P_n = S_n - K(1+r)^{n-N},$$

where C_n (respectively, P_n) is the value of the call option (respectively, put option) on the date n.

EXERCISE 6.3.– Consider the following market model on $\Omega = \{\omega_1, \omega_2, \omega_3, \omega_4\}$:

	$n = 0 : (S_0^0, S_0^1)$	$n = 1 : (S_1^0, S_1^1)$	$n = 2 : (S_2^0, S_2^1)$
ω_1	$(1, 1)$	$(1.1, 0.8)$	$(1.21, 0.64)$
ω_2	$(1, 1)$	$(1.1, 0.8)$	$(1.21, 0.96)$
ω_3	$(1, 1)$	$(1.1, 1.2)$	$(1.21, 0.96)$
ω_4	$(1, 1)$	$(1.1, 1.2)$	$(1.21, 1.44)$

1) Show that the market follows a Cox, Ross and Rubinstein model whose parameters, u and d, are defined.

2) Is this market viable? Complete? If yes, calculate the risk-neutral probability.

Consider a European call option with the strike price $K = 0.9$ and maturity date $N = 2$.

3) For each ω_i (hazard), calculate the profit corresponding to this option. For which hazards will the option be exercised?

4) Calculate the value of the option at each instant.

Consider a European put option with the same strike price $K = 0.9$ and maturity date $N = 2$.

5) For each ω_i, calculate the profit corresponding to this option. For which risks will the options be exercised?

6) Calculate the value of the option at each instant.

EXERCISE 6.4.– Consider the option $X_N = (S_N - K)^2$ in a complete and viable Cox, Ross and Rubinstein market. Calculate the value of this option at each instant.

6.6. Practical work: Simulating the value of a call option

We propose simulating European call options.

1) Write a function R which takes, as argument, a real value x, an integer n and the parameters r, u, d, K and N and which calculates the function c of the price of a European call option at the point (n, x).

2) On the same graph, trace the temporal evolution of the call option C_n and of $(S_n^1 - K)_+$. We will take the parameters $d = -2\%$, $u = 10\%$, $r = 4\%$, $K = 5$ and $N = 50$.

3) What do you observe in the values of C_0 in the different simulations? What do you think of this?

7

American Options

We will now look at a second family of options called *American options*. The chief difference between these and European options lies in the date of exercise: American options can be exercised at any point of time between the initial date and the maturity date. Section 7.1 defines American options and provides a recursive formula to calculate their value. Section 7.2 offers a generic study of the recursive problem posed earlier. Section 7.3 applies the general theory to specific case of American options. Next, section 7.4 examines the particular case of the Cox, Ross and Rubinstein model. Finally, sections 7.5 and 7.6 offer exercises and practical work, respectively.

It is still assumed, throughout this chapter, that Ω has a finite cardinal and that $\mathbb{P}(\omega) > 0$ for any $\omega \in \Omega$. Let N be a time horizon and let $(\mathcal{F}_n)_{0 \leq n \leq N}$ be a filtration such that $\mathcal{F}_0 = \{\emptyset, \Omega\}$ and $\mathcal{F}_N = \mathcal{F}$. In all that follows, it is assumed that the market is viable and complete and $\mathbb{P}^*$ denotes the unique, risk-neutral probability.

7.1. Definition

Let us start by defining an American option.

DEFINITION 7.1.– *An **American option** is a contract that allows the holder to buy or sell a share in the underlying risky asset at a pre-defined strike price K and to do this at any instant between the commencement of the contract and the maturity date N.*

In the following text, Z_n denotes the (*payoff*) generated by the exercising of the option at the instant n, and U_n is the value of the option at the instant n, for $0 \leq n \leq N$. The payoff and the value of the option are processes adapted with respect to the underlying filtration $(\mathcal{F}_n)_{0 \leq n \leq N}$.

EXAMPLE 7.1.– *An American call option with the maturity date N, underlying asset $S = (S_n)_{0 \leq n \leq N}$, and strike price K has the payoff $Z_n = (S_n - K)_+$; an American put option with the same characteristics has the payoff $Z_n = (K - S_n)_+$.* ◇

We now wish to establish the value of an American option. As in the case of European options, the generic formula brings in a conditional expectation. However, the formula obtained here is more complex as there are many more exercise dates possible.

PROPOSITION 7.1.– *We have the backward induction relation*

$$\begin{cases} U_N = Z_N, \\ U_{n-1} = \max\{Z_{n-1}, \frac{1}{1+r}\mathbb{E}^*[U_n|\mathcal{F}_{n-1}]\}, \quad 1 \leq n \leq N. \end{cases}$$

PROOF.– At N, we can only exercise the option instantaneously and obtain Z_N.

At $n - 1 < N$, we either exercise the obtain and obtain the payoff Z_{n-1}, or we do not exercise the option and then have the value $n - 1$ for an admissible strategy with the value U_n at time n. Using the property of the martingale under a risk-neutral probability, we thus have

$$\begin{aligned} V_{n-1}(\Phi) &= (1+r)^{1-n}\widetilde{V}_{n-1}(\Phi) \\ &= (1+r)^{1-n}\mathbb{E}^*[\widetilde{U}_n|\mathcal{F}_{n-1}] \\ &= \frac{(1+r)^{1-n}}{(1+r)^n}\mathbb{E}^*[U_n|\mathcal{F}_{n-1}] \\ &= \frac{1}{1+r}\mathbb{E}^*[U_n|\mathcal{F}_{n-1}]. \end{aligned}$$

The value of the option U_{n-1} is therefore the maximum of the two possibilities, and hence, we have the stated formula. □

We now wish to solve this recurrence relation. We will make use of martingales (and more precisely, supermartingales) once again.

PROPOSITION 7.2.– *The sequence of discounted values $(\widetilde{U}_n)_{0 \leq n \leq N}$ is a supermartingale under $\mathbb{P}^*$. It is the smallest supermartingale dominating the discounted payoffs $(\widetilde{Z}_n)_{0 \leq n \leq N}$.*

PROOF.– Let us show that it is a supermartingale:

The process $(\widetilde{U}_n)$ is well adapted with respect to the underlying filtration.

We are on a finite set Ω; therefore, all the random variables are automatically integrable.

By construction, $\widetilde{U}_{n-1} \geq \mathbb{E}^*[\widetilde{U}_n|\mathcal{F}_{n-1}]$ for any n, and we thus have a supermartingale:

– Let us show that this supermartingale dominates $(\widetilde{Z}_n)$ coordinate by coordinate.

By construction, $\widetilde{U}_n \geq \widetilde{Z}_n$ for any n.

– Finally, let us show that this supermartingale is the smallest dominating the payoffs.

Let $(\widetilde{U}'_n)$ be another supermartingale that dominates $(\widetilde{Z}_n)$. We carry out a demonstration through descending recurrence on n.

For $n = N$, $\widetilde{U}_N = \widetilde{Z}_N \leq \widetilde{U}'_N$.

It is assumed that $\widetilde{U}_{n+1} \leq \widetilde{U}'_{n+1}$ for a fixed n. Then, for any $\omega \in \Omega$, we have

– either $\widetilde{U}_n(\omega) = \widetilde{Z}_n(\omega) \geq \widetilde{U}'_n(\omega)$;

– or $\widetilde{U}_n(\omega) = \mathbb{E}^*[\widetilde{U}_{n+1}|\mathcal{F}_n](\omega) \leq \mathbb{E}^*[\widetilde{U}'_{n+1}|\mathcal{F}_n](\omega) \leq \widetilde{U}'_n$ using the hypothesis of recurrence and the fact that $(\widetilde{U}'_n)$ is a supermartingale.

We thus have $\widetilde{U}'_n \geq \widetilde{U}_n$ for any n. □

The smallest supermartingale that dominates an adapted sequence is called it *Snell envelope*. To study American options in greater detail, we must first study the Snell envelopes, which we do in the next section.

7.2. Optimal stopping

We will now study the optimal stopping problem for an adapted sequence in a generic manner. This will enable us to both calculate the value of an American option, as well as find the optimal dates for exercising the option. We will then return to the application of American options.

7.2.1. *Snell envelope and stopping time*

Let us start with a few definitions.

DEFINITION 7.2.– *Let $(X_n)_{0\leq n\leq N}$ be a sequence adapted with respect to the filtration $(\mathcal{F}_n)_{0\leq n\leq N}$. The smallest supermartingale dominating $(X_n)_{0\leq n\leq N}$ is called the* ***Snell envelope*** *of $(X_n)_{0\leq n\leq N}$. It is denoted by $(M_n)_{0\leq n\leq N}$ and is defined by backward induction through*

$$\begin{cases} M_N = X_N, \\ M_{n-1} = \max\{X_{n-1}, \mathbb{E}^*[M_n|\mathcal{F}_{n-1}]\}, \quad 1 \leq n \leq N. \end{cases}$$

The second part of the definition, the recursive form of the Snell envelope, is demonstrated exactly as was done for Proposition 7.2.

Let $\mathcal{T}$ be the set of stopping times for the filtration $(\mathcal{F}_n)_{0\leq n\leq N}$. Since Ω is finite, this set is also finite. Furthermore, all the stopping times of $\mathcal{T}$ are bounded by N, thus a.s. finite.

DEFINITION 7.3.– *Let $(X_n)_{0\leq n\leq N}$ be a sequence adapted with respect to the filtration $(\mathcal{F}_n)_{0\leq n\leq N}$. An* ***optimal stopping problem*** *consists of finding a stopping time $T^* \in \mathcal{T}$ such that*

$$\mathbb{E}[X_{T^*}] = \sup_{T\in\mathcal{T}} \mathbb{E}[X_T].$$

T^* is then said to be an ***optimal stopping time***

Let us now establish a relationship between the Snell envelope and optimal stopping problem. In all that follows, we take $(X_n)_{0\leq n\leq N}$, a sequence adapted with respect to the filtration $(\mathcal{F}_n)_{0\leq n\leq N}$ and $(M_n)_{0\leq n\leq N}$, its Snell envelope.

PROPOSITION 7.3.– *A stopping time $T^* \leq N$ is an optimal stopping time if and only if $X_{T^*} = M_{T^*}$ and $(M_n^{T^*})$ is a martingale.*

PROOF.– If these two conditions are satisfied, then, on the one hand, we have

$$M_0 = M_0^{T^*} = \mathbb{E}[M_N^{T^*}|\mathcal{F}_0] = \mathbb{E}[M_N^{T^*}] = \mathbb{E}[M_{T^*}] = \mathbb{E}[X_{T^*}],$$

because $T^* \leq N$ and $(M_n^{T^*})$ is a martingale. Furthermore, for any $T \in \mathcal{T}$, we have the fact that (M_n^T) is a supermartingale, since the Snell envelope is a supermartingale, and any stopped supermartingale is a supermartingale. Thus, we have

$$M_0 = M_0^T \geq \mathbb{E}[M_N^T|\mathcal{F}_0] = \mathbb{E}[M_N^T] = \mathbb{E}[M_T] \geq \mathbb{E}[X_T],$$

because M is greater than X. Therefore, we obtained that $\mathbb{E}[X_{T^*}] = M_0 \geq \mathbb{E}[X_T]$ for any $T \in \mathcal{T}$, T^* is therefore an optimal stopping time.

In passing, we have obtained that $M_0 = \sup_{T\in\mathcal{T}} \mathbb{E}[X_T]$, which will be useful further on in the text.

Conversely, if T^* is an optimal stopping time, since it is a stopping time, we still have, according to the supermartingale property and the fact that M dominates X

$$M_0 = M_0^{T^*} \geq \mathbb{E}[M_N^{T^*}|\mathcal{F}_0] = \mathbb{E}[M_N^{T^*}] = \mathbb{E}[M_{T^*}] \geq \mathbb{E}[X_{T^*}] = M_0,$$

since T^* is optimal. The inequalities in this sequence are, therefore, equalities. In particular, it follows that $\mathbb{E}[M_{T^*}] = \mathbb{E}[X_{T^*}]$. Furthermore, it is known that $M_{T^*} \geq X_{T^*}$ as the Snell envelope dominates the sequence; therefore, $M_{T^*} = X_{T^*}$.

By also introducing an intermediate time n, the same arguments give

$$M_0 = M_0^{T^*} \geq \mathbb{E}[M_n^{T^*}|\mathcal{F}_0] \geq \mathbb{E}\big[\mathbb{E}[M_N^{T^*}|\mathcal{F}_n]|\mathcal{F}_0\big] \geq \mathbb{E}[M_N^{T^*}|\mathcal{F}_0] = M_0,$$

thus, we once again have an equality and since $\mathbb{E}[M_N^{T^*}|\mathcal{F}_n] \leq M_n^{T^*}$ using the supermartingale property, we do indeed have the fact that $\mathbb{E}[M_N^{T^*}|\mathcal{F}_n] = M_n^{T^*}$; therefore, we have a martingale. □

7.2.2. *Construction of optimal stopping times*

We have just seen that it is possible to characterize optimal stopping times with the Snell envelope. We now wish to explicitize the optimal stopping times. Proposition 7.3 gives us at least two paths for this:

– finding the stopping times that equalize X and M;

– finding the stopping times that transform the stopped supermartingale into a martingale.

A natural way to proceed with the first path is to introduce the first instant where $X_n = M_n$. A method to proceed with the second path is to look at the Doob decomposition of the supermartingale (M_n) (see section 4.3), and to work so that its increasing process remains zero. It will be seen that both these paths do indeed yield optimal stopping times.

PROPOSITION 7.4.– *We write* $T_0 = \inf\{n; M_n = X_n\}$. *Therefore,* T_0 *is an optimal stopping time.*

PROOF.– Let us start by showing that T_0 is indeed an optimal stopping time $\mathcal{T}$.

T_0 is bounded by N because $M_N = X_N$. Furthermore,

$$(T_0 = n) = (X_0 < M_0, X_1 < M_1, \dots X_{n-1} < M_{n-1}, X_n = M_n) \in \mathcal{F}_n,$$

therefore, this is indeed a stopping time.

Let us now show that it is optimal, by using the characterization given in Proposition 7.3. On the one hand, from the definition of T_0, we directly obtain $X_{T_0} = M_{T_0}$.

On the other hand, since T_0 is a stopping time and (M_n) is a supermartingale, it is known that $(M_n^{T_0})$ is again a supermartingale. In addition, we have the decomposition

$$\begin{aligned}
M_{n+1}^{T_0} &= M_{n+1}^{T_0}\mathbb{1}_{(T_0 \geq n+1)} + M_{n+1}^{T_0}\mathbb{1}_{(T_0 < n+1)} \\
&= M_{n+1}\mathbb{1}_{(T_0 \geq n+1)} + M_{T_0}\mathbb{1}_{(T_0 < n+1)} \\
&= M_{n+1}\mathbb{1}_{(T_0 \geq n+1)} + M_n^{T_0}\mathbb{1}_{(T_0 < n+1)} \\
&= M_n^{T_0} - M_n^{T_0}\mathbb{1}_{(T_0 \geq n+1)} + M_{n+1}\mathbb{1}_{(T_0 \geq n+1)} \\
&= M_n^{T_0} + \mathbb{1}_{(T_0 \geq n+1)}(M_{n+1} - M_n).
\end{aligned}$$

Thus, we have

$$\begin{aligned}
\mathbb{E}[M_{n+1}^{T_0}|\mathcal{F}_n] &= \mathbb{E}[M_n^{T_0} + \mathbb{1}_{(T_0 \geq n+1)}(M_{n+1} - M_n)|\mathcal{F}_n] \\
&= M_n^{T_0} + \mathbb{1}_{(T_0 \geq n+1)}\mathbb{E}[M_{n+1} - M_n|\mathcal{F}_n]
\end{aligned}$$

since $(T_0 \geq n+1) = (T_0 \leq n)^c \in \mathcal{F}_n$. However, on the event $(T_0 \geq n+1)$, we have $M_n > X_n$ because T_0 is the first instant where these two quantities are equal and because the Snell envelope dominates the process. Thus, still staying on this event, from the definition of the Snell envelope, it follows that

$$M_n = \max\{X_n, \mathbb{E}^*[M_{n+1}|\mathcal{F}_n]\} = \mathbb{E}[M_{n+1}|\mathcal{F}_n].$$

We thus obtain that $\mathbb{E}[M_{n+1}^{T_0}|\mathcal{F}_n] = M_n^{T_0}$ and we do indeed have a martingale.

Therefore, using Proposition 7.3, T_0 is indeed an optimal stopping time. □

Let us now introduce the Doob decomposition of the supermartingale (M_n):

$$M_n = M_0 + \mathcal{M}_n - \mathcal{A}_n,$$

with $\mathcal{M}_0 = \mathcal{A}_0 = 0$, $(\mathcal{M}_n)$ is a martingale and $(\mathcal{A}_n)$ is a predictable increasing process.

PROPOSITION 7.5.– *Let $\overline{T}$ be defined by $\overline{T} = N \wedge \inf\{n; \mathcal{A}_{n+1} \neq 0\}$. Then, $\overline{T}$ is an optimal stopping time.*

PROOF.– Let us start by showing that $\overline{T}$ is a stopping time. We have

$$(\overline{T} = N) = (\mathcal{A}_n = 0, \forall 0 \leq n \leq N) \in \mathcal{F}_N,$$

and for $n < N$,

$$(\overline{T} = n) = (\mathcal{A}_k = 0, \forall 0 \leq k \leq n, \mathcal{A}_{n+1} > 0) \in \mathcal{F}_n,$$

since $\mathcal{A}$ is predictable. We thus have a stopping time.

Let us now show that it is optimal by once again using the characterization from Proposition 7.3. On the one hand, by definition, $\mathcal{A}_{\overline{T}} = 0$ and for any $n \leq \overline{T}$, $\mathcal{A}_n = 0$ by monotonicity; therefore, $(M_n^{\overline{T}}) = (\mathcal{M}_n^{\overline{T}})$ is a martingale.

On the other hand, from the definition of the Snell envelope, we have

$$\begin{aligned} M_{\overline{T}} &= \sum_{n=0}^{N-1} \mathbb{1}_{(\overline{T}=n)} M_n + \mathbb{1}_{(\overline{T}=N)} M_N \\ &= \sum_{n=0}^{N-1} \mathbb{1}_{(\overline{T}=n)} \max\{X_n, \mathbb{E}[M_{n+1}|\mathcal{F}_n]\} + \mathbb{1}_{(\overline{T}=N)} X_N. \end{aligned}$$

Furthermore, on the event $(T = n < N)$, we have $M_n = \mathcal{M}_n = \mathbb{E}[\mathcal{M}_{n+1}|\mathcal{F}_n] < \mathbb{E}[M_{n+1}|\mathcal{F}_n]$ because $\mathcal{A}_{n+1} > 0$. Therefore, on $(T = n < N)$, $M_n = X_n$. Thus, we have

$$\begin{aligned} M_{\overline{T}} &= \sum_{n=0}^{N-1} \mathbb{1}_{(\overline{T}=n)} \max\{X_n, \mathbb{E}[M_{n+1}|\mathcal{F}_n]\} + \mathbb{1}_{(\overline{T}=N)} X_N \\ &= \sum_{n=0}^{N-1} \mathbb{1}_{(\overline{T}=n)} X_n + \mathbb{1}_{(\overline{T}=N)} X_N \\ &= X_{\overline{T}}. \end{aligned}$$

Therefore, $\overline{T}$ is indeed an optimal stopping time. □

By construction, T_0 is the smallest of the optimal stopping times, because they must all satisfy $X_{T^*} = M_{T^*}$. We will now see that $\overline{T}$ is the largest of the optimal stopping times.

PROPOSITION 7.6.– *The stopping time $\overline{T}$ is the largest optimal stopping time.*

PROOF.– It is known that $\overline{T}$ is indeed an optimal stopping time. Reasoning by the absurd, let us assume that T^* is an optimal stopping time such that $\mathbb{P}(T^* > \overline{T}) > 0$. Thus, it is possible to write

$$M_{T^*} = \mathbb{E}[M_0 + \mathcal{M}_{T^*} - \mathcal{A}_{T^*}] = \mathbb{E}[M_0] + \mathbb{E}[\mathcal{M}_0] - \mathbb{E}[\mathcal{A}_{T^*}] = \mathbb{E}[M_0] - \mathbb{E}[\mathcal{A}_{T^*}]$$

from the Doob stopping theorem ($T^* \leq N$) using the fact that $(\mathcal{M}_n)$ is a martingale starting from 0. Furthermore, by construction of $\overline{T}$, on the event $(T^* \leq \overline{T})$, we have $\mathcal{A}_{T^*} = 0$, and on $(T^* > \overline{T})$, we have $\mathcal{A}_{T^*} > 0$. Since $\mathbb{P}(T^* > \overline{T}) > 0$, it can be derived from this that $\mathbb{E}[\mathcal{A}_{T^*}] > 0$. Therefore, $M_{T^*} < \mathbb{E}[M_0]$ which contradicts the fact that $(M_n^{T^*})$ is a martingale via the Doob stopping theorem. Thus, we do indeed have $T^* \leq \overline{T}$ almost surely. □

7.3. Application to American options

Let us now apply the optimal stopping theory to American options. It has been seen that the discounted value $(\widetilde{U}_n)$ of the American option with payoff (Z_n) is the Snell envelope of $(\widetilde{Z}_n)$.

THEOREM 7.1 (Hedging American options).– *The seller of an American option may perfectly hedge their position: there exists an admissible strategy Φ such that $V_n(\Phi) \geq U_n$ for any n.*

PROOF.– The Doob decomposition of $(\widetilde{U}_n)$ is $\widetilde{U}_n = \widetilde{U}_0 + \widetilde{\mathcal{M}}_n - \widetilde{\mathcal{A}}_n$, $0 \leq n \leq N$, where $(\widetilde{\mathcal{M}}_n)$ is a martingale under $\mathbb{P}^*$ and $(\widetilde{\mathcal{A}}_n)$ is predictable and increasing. Since the market is complete, there exists an admissible strategy Φ such that $\widetilde{V}_N(\Phi) = \widetilde{\mathcal{M}}_N$. Since $(\widetilde{V}_n(\Phi)$ is a martingale under $\mathbb{P}^*$, we have

$$\widetilde{V}_n(\Phi) = \mathbb{E}^*[\widetilde{V}_N(\Phi)|\mathcal{F}_n] = \mathbb{E}^*[\widetilde{\mathcal{M}}_N|\mathcal{F}_n] = \widetilde{\mathcal{M}}_n = \widetilde{U}_n + \widetilde{\mathcal{A}}_n \geq \widetilde{U}_n.$$

therefore, following the strategy Φ, we can indeed hedge for the option. □

THEOREM 7.2 (Date of exercising American options).– *The optimal dates for exercising an American option are the optimal stopping times.*

PROOF.– An exercise date cannot anticipate the future; therefore, it is indeed a stopping time.

We then use the characterization from Proposition 7.3.

– Let T be an exercise date. If $U_T > Z_T$, the exercise is not recommended as we lose an option with the value U_T against a payoff $Z_T < U_T$. Therefore, on the optimal exercise date, we must have $Z_T = U_T$.

– At $\overline{T}$, it is possible to sell the option against a wealth of $\widetilde{U}_{\overline{T}} = \widetilde{V}_{\overline{T}}(\Phi)$ (because $\widetilde{\mathcal{A}}_{\overline{T}} = 0$), and follow the hedging strategy Φ. At any date later than $\overline{T}$, we will then have $\widetilde{V}_n(\Phi) > \widetilde{U}_n$, since $\widetilde{\mathcal{A}}_n > 0$. Therefore, an optimal exercise date must be less than or equal to $\overline{T}$. Finally, if $T \leq \overline{T}$, $(\widetilde{U}_n^T)$ is a martingale because $\widetilde{\mathcal{A}}_T = 0$.

Thus, an optimal exercise date is indeed an optimal stopping time. □

Based on the above section, it is possible to thus construct at least two optimal stopping times: T_0 and $\overline{T}$.

7.4. The Cox, Ross and Rubinstein model

Let us now apply the above generic results to the specific case of a Cox, Ross and Rubinstein market.

We work on a viable and complete Cox, Ross and Rubinstein market, that is, a market that satisfies $d < r < u$, and we work under a risk-neutral probability characterized by $p = \frac{r-d}{u-d}$.

7.4.1. *Value of American options*

Let us start by explicitizing the value of an American option at each instant.

PROPOSITION 7.7.– *The value of the American put option with underlying asset S^1, maturity date N and strike price K is $P_n = f(n, S_n^1)$, where f is the deterministic function defined through backward induction by*

$$\begin{cases} f(N,x) = (K-x)_+ \\ f(n,x) = \max\left\{(K-x)_+; \dfrac{(1-p)f(n+1,x(1+d)) + p)f(n+1,x(1+u))}{1+r}\right\}, \\ 0 \leq n \leq N-1. \end{cases}$$

PROOF.– By backward induction on n.

For $n = N$, we have $P_N = Z_N = (K - S_N^1)_+$.

Let us assume the formula is true for step $n+1$ for a fixed n. We then have

$$P_n = \max\{Z_n; \frac{1}{1+r}\mathbb{E}^*[P_{n+1}|\mathcal{F}_n]\}.$$

Furthermore, we have

$$\begin{aligned} \mathbb{E}^*[[P_{n+1}|\mathcal{F}_n]\} &= \mathbb{E}^*[f(n+1, S_{n+1}^1|\mathcal{F}_n]\} \\ &= \mathbb{E}^*[f(n+1, S_n^1 T_{n+1}|\mathcal{F}_n]\} \\ &= (1-p)f(n+1, S_n^1(1+d)) + p)f(n+1, S_n^1(1+u)), \end{aligned}$$

using the independence of the T_n under $\mathbb{P}^*$, hence the result. □

Unlike in the case of European options, there is no simple closed-form formula to resolve the recursion.

7.4.2. *Hedging American options*

We now wish to construct an explicit hedging strategy.

PROPOSITION 7.8.– *It is possible to explicitly construct an admissible strategy that enables us to hedge with an American put option with the underlying asset* S^1, *maturity date* N *and strike price* K. *The number of assets to be held between the instant* n *and the instant* $n+1$ *are*

$$\phi^0_{n+1} = \Delta^0(n+1, S^1_n)$$
$$\phi^1_{n+1} = \Delta^1(n+1, S^1_n)$$

where Δ^0 *and* Δ^1 *are the deterministic functions defined by*

$$\Delta^0(n,x) = \frac{(1+u)f(n,x(1+d)) - (1+d)f(n,x(1+u))}{(1+r)^n(u-d)}$$

$$\Delta^1(n,x) = \frac{f(n,x(1+d)) - f(n,x(1+u))}{x(u-d)}.$$

PROOF.– The proof is identical to that used in Proposition 6.4 and is thus omitted here. □

7.5. Exercises

EXERCISE 7.1.– Let $\widetilde{C}^a_n$ be the discounted price at the instant n of an American option with the payoff (Z_n), and let $\widetilde{C}^e_n$ be the discounted price at the instant n of the European option with the same payoff Z_N and the same maturity date N.

1) Show by backward induction on n that $\widetilde{C}^a_n \geq \widetilde{C}^e_n$ for any $0 \leq n \leq N$.

2) If $\widetilde{C}^e_n \geq \widetilde{Z}_n$ for any $0 \leq n \leq N$, show that $\widetilde{C}^a_n = \widetilde{C}^e_n$ for any $0 \leq n \leq N$.

3) Deduce the equality at each instant of the value of a European call option and an American call option with the same characteristics.

EXERCISE 7.2.– Consider an American call option with the maturity date N and strike price K in a market that follows the Cox, Ross and Rubinstein model. The value of the option at the time n is denoted by C_n. Show through backward induction on n that $C_n = g(n, S^1_n)$ with

$$\begin{cases} g(N,x) = (x-K)_+, \\ g(n,x) = \max\left\{(x-K)_+;\ \frac{pg\left(n+1,x(1+u)\right)+(1-p)g\left(n+1,x(1+d)\right)}{1+r}\right\}. \end{cases}$$

EXERCISE 7.3.– Consider a financial market that follows the Cox, Ross and Rubinstein model with parameters $r = 0$, $d = -0.1$, $u = 0.1$ on $\Omega = \{\omega_1, \dots, \omega_8\}$. The risky asset takes the following values:

	$n = 0$	$n = 1$	$n = 2$	$n = 3$
ω_1	1	1.1	1.21	1.331
ω_2	1	1.1	1.21	1.089
ω_3	1	1.1	0.99	1.089
ω_4	1	1.1	0.99	0.891
ω_5	1	0.9	0.99	1.089
ω_6	1	0.9	0.99	0.891
ω_7	1	0.9	0.81	0.891
ω_8	1	0.9	0.81	0.729

1) Is this market viable? Complete? If yes, calculate the parameter p of the unique risk-neutral probability.

Consider an American call option on this market with the strike price $K = 0.9$ and maturity date $N = 3$.

2) For each instant and for each ω_i, calculate the payoff resulting from the option being exercised.

3) For each instant and for each ω_i, calculate the value of the option.

4) For each ω_i, give an optimal exercise date for the option.

Consider an American put option on this market with the strike price $K = 0.9$ and maturity date $N = 3$.

5) For each instant and each ω_i, calculate the payoff of the option.

6) For each instant and for each ω_i, calculate the value of the option.

7) For each ω_i, give an optimal exercise date for the option.

7.6. Practical work

We will now illustrate the calculation of American put options. To simplify the simulations, it is assumed in all that follows that $r = 0$, thus, the processes are equal to the discounted processes.

1) Write a function R that takes as argument the parameters u and d, a strike price K and a horizon N and that produces a trajectory $(Z_n)_{0 \leq n \leq N}$ for the gain corresponding to the exercise of the underlying American option (S_n) following the Cox, Ross and Rubinstein model with strike price K and maturity date N.

2) Graphically represent a few trajectories for $u = 10\%$, $d = -5\%$, $K = 1.2$ and $N = 50$.

3) Write a recursive function R that takes as argument a real value x, an integer n and the parameters u, d, K and N and that calculates the function f of the price of an American put option at the point (n, x).

4) Write a function R that takes as argument the parameters u and d, a strike price K and a horizon N and that produces a trajectory $(U_n, Z_n)_{0 \leq n \leq N}$ for the gain corresponding to the exercise of, and to the value of, the underlying American value (S_n) following the Cox, Ross and Rubinstein model, with strike price K and maturity date N.

5) Graphically represent a few trajectories of (U, Z) for $u = 10\%$, $d = -5\%$, $K = 1.2$ and $N = 20$.

6) For each of the above graphs, determine the value of the optimal stopping time T_0.

7) What do you observe regarding the values of U_0 in your different simulations? What do you think about this?

8

Solutions to Exercises and Practical Work

8.1. Solutions to exercises in Chapter 1

8.1.1. *Exercise 1.1*

1) Let us describe all the σ-algebras of $\Omega = \{a, b, c\}$ based on their cardinal. The smallest is $\mathcal{F}_0 = \{\emptyset, \Omega\}$. Those generated by an element are

$$\sigma(\{a\}) = \{\emptyset, \{a\}, \{b, c\}, \Omega\},$$
$$\sigma(\{b\}) = \{\emptyset, \{b\}, \{a, c\}, \Omega\},$$
$$\sigma(\{c\}) = \{\emptyset, \{c\}, \{a, b\}, \Omega\}.$$

The largest σ-algebra is that generated by two elements that corresponds to the set of subsets of Ω:

$$\mathcal{P}(\Omega) = \{\emptyset, \{a\}, \{b\}, \{c\}, \{a, b\}, \{a, c\}, \{b, c\}, \Omega\}.$$

2) The only inclusions are

$$\mathcal{F}_0 \subset \sigma(\{a\}) \subset \mathcal{P}(\Omega),$$
$$\mathcal{F}_0 \subset \sigma(\{b\}) \subset \mathcal{P}(\Omega),$$
$$\mathcal{F}_0 \subset \sigma(\{c\}) \subset \mathcal{P}(\Omega).$$

The trivial σ-algebra is contained in all the σ-algebras and all the σ-algebras are sub-σ-algebras of $\mathcal{P}(\Omega)$.

8.1.2. *Exercise 1.2*

1) The family $\mathcal{A}_1 = \{\emptyset, \Omega, \{b\}, \{a, c, d\}\}$ is indeed a σ-algebra: it contains Ω, it is stable under complement and it is stable under countable union.

2) The family $\mathcal{A}_2 = \{\emptyset, \Omega, \{b,c,d\}, \{c,d\}\}$ is not a σ-algebra. For example, it does not contain the complement $\{a\}$ of $\{b,c,d\}$. The σ-algebra that it generates is

$$\sigma(\mathcal{A}_2) = \{\emptyset, \{a\}, \{c,d\}, \{b,c,d\}, \{a,b\}, \{a,c,d\}, \{b\}, \Omega\}.$$

3) The family $\mathcal{A}_3 = \{\{b\}\}$ is not a σ-algebra. For example, it does not contain $\emptyset$. The σ-algebra that it generates is

$$\sigma(\mathcal{A}_3) = \{\emptyset, \Omega, \{b\}, \{a,c,d\}\} = \mathcal{A}_1.$$

4) The family $\mathcal{A}_4 = \{\emptyset, \Omega, \{a,c\}, \{b,d\}\}$ is indeed a σ-algebra: it contains Ω, it is stable under complement and it is stable under countable union.

8.1.3. *Exercise 1.3*

To show that X is $\mathcal{G}$-measurable if and only if $\sigma(X) \subset \mathcal{G}$, we proceed through double implication.

– If X is $\mathcal{G}$-measurable, then by definition, for any $A \in \sigma(X) = \sigma(X^{-1}(\mathcal{E}))$, we have $A \in \mathcal{G}$, therefore $\sigma(X) \subset \mathcal{G}$.

– If $\sigma(X) \subset \mathcal{G}$, then for any $A \in \sigma(X) = \sigma(X^{-1}(\mathcal{E}))$, we have $A \in \mathcal{G}$, therefore X is $\mathcal{G}$-measurable.

8.1.4. *Exercise 1.4*

We wish to show that $\mathbb{1}_A$ is $\mathcal{G}$-measurable if and only if $A \in \mathcal{G}$. In order to do this, we will completely describe the σ-algebra $\sigma(\mathbb{1}_A)$. Let B be a Borel set of $\mathbb{R}$. Describe the event $(\mathbb{1}_A \in B)$. We have

$$(\mathbb{1}_A \in B) = \begin{cases} \emptyset & \text{if } 0 \notin B \text{ and } 1 \notin B, \\ A & \text{if } 0 \notin B \text{ and } 1 \in B, \\ A^c & \text{if } 0 \in B \text{ and } 1 \notin B, \\ \Omega & \text{if } 0 \in B \text{ and } 1 \in B. \end{cases}$$

Therefore, the σ-algebra generated by $\mathbb{1}_A$ is

$$\sigma(\mathbb{1}_A) = \sigma(A) = \{\emptyset, A, A^c, \Omega\},$$

and $\mathbb{1}_A$ is $\mathcal{G}$-measurable if and only if $\sigma(\mathbb{1}_A) \subset \mathcal{G}$ as per Exercise 1.3, therefore, if and only if $A \subset \mathcal{G}$.

8.1.5. *Exercise 1.5*

Let $\Omega = \{P, F\} \times \{P, F\}$ and $\mathcal{F} = \mathcal{P}(\Omega)$, corresponding to two successive coin tosses using the same coin. Let

– X_1 be the random variable *the number of T on the first toss;*

– X_2 be the *number of T on the second toss;*

– Y be the *number of T obtained on both tosses;*

– and $Z = 1$ if both tosses yield an identical result, if not, $Z = 0$.

1) The random variable X_1 is the function: $\Omega \to \{0, 1\}$ defined by

$$X_1(\omega_1, \omega_2) = \mathbb{1}_P(\omega_1),$$

and

$$(X_1 = 0) = \{FF, FP\}$$
$$(X_1 = 1) = \{PP, PF\}.$$

Thus, the σ-algebra generated by X_1 is

$$\begin{aligned}\mathcal{F}_1 &= \sigma(\{FF, FP\}, \{PP, PF\}) \\ &= \{\emptyset, \Omega, \{PP, PF\}, \{FP, FF\}\}.\end{aligned}$$

Similarly, X_2 is the function: $\Omega \to \{0, 1\}$ defined by

$$X_2(\omega_1, \omega_2) = \mathbb{1}_P(\omega_2),$$

and $\mathcal{F}_2 = \{\emptyset, \Omega, \{PP, FP\}, \{FF, PF\}\}$.

In particular, $\mathcal{F}_1 \not\subset \mathcal{F}_2$ therefore X_1 is not $\mathcal{F}_2$-measurable.

2) The random variable Y is the function: $\Omega \to \{0, 1, 2\}$ defined by

$$Y(\omega_1, \omega_2) = \mathbb{1}_P(\omega_1) + \mathbb{1}_P(\omega_2),$$

and

$$(Y = 0) = \{FF\}$$
$$(Y = 1) = \{FP, PF\}$$
$$(Y = 2) = \{FF\}.$$

Thus, the σ-algebra generated by Y is

$$\mathcal{G}_1 = \{\emptyset, \Omega, \{FF\}, \{PF, FP\}, \{PP\}, \{PF, FP, PP\}, \{PP, FF\}, \{FF, PF, FP\}\}.$$

It can be seen that $\mathcal{G} \not\subset \mathcal{F}_1$, thus Y is not $\mathcal{F}_1$-measurable. Further, $\mathcal{F}_1 \not\subset \mathcal{G}$, therefore X_1 is not $\mathcal{G}$-measurable.

3) The random variable Z is the function: $\Omega \to \{0, 1\}$ defined by

$$Z(\omega_1, \omega_2) = \mathbb{1}_{(\omega_1 = \omega_2)},$$

and

$$(Z = 0) = \{FP, PF\}$$
$$(Z = 1) = \{FF, PP\}.$$

Therefore, the σ-algebra generated by Z is

$$\mathcal{H} = \{\emptyset, \Omega, \{PP, FF\}, \{FP, PF\}\}.$$

It can be seen that $\mathcal{H} \not\subset \mathcal{F}_1$, therefore Z is not $\mathcal{F}_1$-measurable. On the contrary, $\mathcal{H} \subset \mathcal{G}$, therefore Z is $\mathcal{H}$-measurable. Indeed, we have $Z = \mathbb{1}_{\{0,2\}}(Y)$. Conversely, $\mathcal{G} \not\subset \mathcal{H}$, therefore Y is not $\mathcal{H}$-measurable.

4) All the σ-algebras are sub-σ-algebras of $\mathcal{F}$. The only other inclusion is $\mathcal{H} \subset \mathcal{G}$.

8.1.6. *Exercise 1.6*

Let $(X_n)_{n \geq 1}$ be a sequence of independent random variables with the same Rademacher distribution with parameter $1/2$

$$\mathbb{P}(X_1 = 1) = \mathbb{P}(X_1 = -1) = 1/2.$$

Let $S_0 = 0$ and $S_n = \sum_{k=1}^{n} X_k$ for $n \geq 1$. The process $S = (S_n)_{n \in \mathbb{N}}$ is called the simple symmetric random walk on $\mathbb{Z}$. It will be studied in detail in Chapter 3. We write $X_0 = 0$. We wish to show that the filtration generated by the sequence $(X_n)_{n \in \mathbb{N}}$ is the same as that generated by the sequence $(S_n)_{n \in \mathbb{N}}$. We carry out a demonstration by induction on n.

– Initialization: for $n = 0$, we have $S_0 = X_0 = 0$, therefore $\sigma(S_0) = \sigma_(X_0)$.

– Induction hypothesis: let us assume that for a fixed n,

$$\sigma(S_0, \ldots, S_n) = \sigma(X_0, \ldots, X_n).$$

– Heredity: at the rank $n + 1$, we have $S_{n+1} = \sum_{k=1}^{n+1} X_k$ is $\sigma(X_0, \dots, X_{n+1})$-measurable.

Using the induction hypothesis, the S_k, $k \leq n$ are also $\sigma(X_0, \dots, X_n)$-measurable, therefore $\sigma(X_0, \dots, X_{n+1})$-measurable. Thus, we do indeed have

$$\sigma(S_0, \dots, S_{n+1}) \subset \sigma(X_0, \dots, X_{n+1}).$$

Conversely, $X_{n+1} = S_{n+1} - S_n$ is $\sigma(S_0, \dots, S_{n+1})$-measurable.

Using the induction hypothesis, the X_k, $k \leq n$ are also $\sigma(S_0, \dots, S_n)$-measurable, therefore $\sigma(S_0, \dots, S_{n+1})$-measurable.

Thus, we have $\sigma(X_0, \dots, X_{n+1}) \subset \sigma(S_0, \dots, S_{n+1})$. Since we have a double inclusion, we have

$$\sigma(X_0, \dots, X_{n+1}) = \sigma(S_0, \dots, S_{n+1}).$$

8.1.7. *Exercise 1.7*

Consider the following game of chance. A gambler begins by choosing a number between 6 and 8 inclusive, which is called the principal. The gambler rolls two dice (non-loaded, six-sided) and adds up the results. The stakes are as follows:

– if the sum is 2 or 3, the gambler loses 1 DT;

– if the sum is 11, the gambler wins 1 DT if the principal is 7, otherwise they lose 1 DT;

– if the sum is 12, the gambler wins 1 DT if the principal is 6 or 8, otherwise they lose 1 DT;

– finally, in all other cases, nothing happens (no win, no loss).

1) The space of all the outcomes of the experiment where both dice are rolled is $\Omega = \{1, 2, ..., 6\} \times \{1, 2, ..., 6\}$.

2) The possible values of the random variable S, the random variable giving the sum of both the dice are $S(\Omega) = \{2, 3, 4, 5, 6, 7, 8, 9, 10, 11, 12\}$ and its distribution is given in Table 8.1.

k	2	3	4	5	6	7	8	9	10	11	12
$\mathbb{P}(S = k)$	$\frac{1}{36}$	$\frac{2}{36}$	$\frac{3}{36}$	$\frac{4}{36}$	$\frac{5}{36}$	$\frac{6}{36}$	$\frac{5}{36}$	$\frac{4}{36}$	$\frac{3}{36}$	$\frac{2}{36}$	$\frac{1}{36}$

Table 8.1. *Distribution of the random variable S*

3) The possible values of the random variable X_7, giving the gambler's winnings when the principal is 7, are $X_7(\Omega) = \{-1, 0, 1\}$, and the distribution of X_7 is given by

$$\mathbb{P}(X_7 = 1) = \mathbb{P}(S = 7) + \mathbb{P}(S = 11) = \frac{2}{9},$$

$$\mathbb{P}(X_7 = -1) = \mathbb{P}(S = 2) + \mathbb{P}(S = 3) + \mathbb{P}(S = 1) = \frac{1}{9},$$

$$\mathbb{P}(X_7 = 0) = 1 - \mathbb{P}(X_7 = -1) - \mathbb{P}(X_7 = 1) = \frac{2}{3},$$

and is summarized in Table 8.2.

k	-1	0	1
$\mathbb{P}(X_7 = k)$	$\frac{2}{9}$	$\frac{1}{9}$	$\frac{6}{9}$

Table 8.2. *Distribution of the random variable X_7*

4) The expectation of X_7 is given by:

$$\begin{aligned}\mathbb{E}[X_7] &= -1 \times \mathbb{P}(X_7 = -1) + 0 \times \mathbb{P}(X_7 = 0) + 1 \times \mathbb{P}(X_7 = 1) \\ &= \frac{-1}{9} + \frac{2}{9} = \frac{1}{9}.\end{aligned}$$

On the other hand, the second-order moment of X_7 is

$$\begin{aligned}\mathbb{E}[X_7^2] &= (-1)^2 \times \mathbb{P}(X_7 = -1) + 1 \times \mathbb{P}(X_7 = 1) \\ &= \frac{1}{9} + \frac{2}{9} = \frac{1}{3},\end{aligned}$$

and thus its variance is

$$\mathbb{V}(X_7) = \mathbb{E}(X_7^2) - \mathbb{E}(X_7)^2 = \frac{1}{3} - \frac{1}{81} = \frac{26}{81}.$$

k	-1	0	1
$\mathbb{P}(X_6 = k)$	$\frac{5}{36}$	$\frac{25}{36}$	$\frac{6}{36}$

Table 8.3. *The distribution of the random variable* X_6

5) The possible values for the random variable X_6, giving the gambler's winnings when the principal is 6, are $X_6(\Omega) = \{-1, 0, 1\}$, and the distribution of X_6 is given by

$$\mathbb{P}(X_6 = 1) = \mathbb{P}(S = 6) + \mathbb{P}(S = 12) = \frac{1}{6},$$

$$\mathbb{P}(X_6 = -1) = \mathbb{P}(S = 2) + \mathbb{P}(S = 3) + \mathbb{P}(S = 11) = \frac{5}{36},$$

$$\mathbb{P}(X_6 = 0) = 1 - \mathbb{P}(X_6 = -1) - \mathbb{P}(X_6 = 1) = \frac{25}{36},$$

and summarized in Table 8.3. Its expectation is

$$\begin{aligned}\mathbb{E}[X_6] &= -1 \times \mathbb{P}(X_6 = -1) + 0 \times \mathbb{P}(X_6 = 0) + 1 \times \mathbb{P}(X_6 = 1) \\ &= \frac{-5}{36} + \frac{1}{6} = \frac{1}{36}.\end{aligned}$$

8.1.8. *Exercise 1.8*

Let $p \in]0, 1[$. We have a coin that turns up T with the probability p. This coin is tossed until we obtain T for the second time. Let X be the number of H obtained over the course of this experiment.

1) Let us determine the distribution of X. For a fixed $n \in \mathbb{N}$, the occurrence of the event $\{X = n\}$ signifies that we have obtained one and only one T during the first $n + 1$ attempts, and the $n + 2$nd attempt produces T. There are therefore $n + 1$ choices for the first T. With this chosen, the elementary event has a probability of $p^2(1 - p)^n$ for 2 T and n H. The distribution of X is thus given by

$$\mathbb{P}(X = n) = (n + 1)p^2(1 - p)^n,$$

for any $n \in \mathbb{N}$.

2) The series defining the expectation of X has the general term

$$n\mathbb{P}(X = n) = n(n+1)p^2(1-p)^n.$$

As the radius of convergence of the power series $\sum_n n(n+1)z^n$ is equal to 1, and since $|1-p| < 1$, the series

$$\sum_n n\mathbb{P}(X = n) = p^2 \sum_n n(n+1)(1-p)^n$$

is convergent, and consequently the expectation of X exists.

3) We have

$$\mathbb{E}[X] = \sum_{n=1}^{+\infty} n\mathbb{P}(X = n) = p^2 \sum_{n=1}^{+\infty} n(n+1)z^n,$$

with $z = 1-p \in]0,1[$. As each of the power series $\sum_n n(n+1)z^n$, $\sum_n nz^{n+1}$ and $\sum_n z^n$ is absolutely convergent in the open interval $]-1,1[$, we have

$$\begin{aligned}
\sum_{n=1}^{+\infty} n(n+1)z^n &= \left(\sum_{n=1}^{+\infty} nz^{n+1}\right)' \\
&= \left(z^2 \sum_{n=1}^{+\infty} nz^{n-1}\right)' \\
&= \left(z^2 \left[\sum_{n=0}^{+\infty} z^n\right]'\right)' \\
&= \left(z^2 \left[\frac{1}{1-z}\right]'\right)' \\
&= \frac{2z}{(1-z)^3}.
\end{aligned}$$

Therefore, we obtain

$$\mathbb{E}[X] = p^2 \sum_{n=1}^{+\infty} n(n+1)(1-p)^n = \frac{2(1-p)}{p}.$$

8.2. Solutions to exercises in Chapter 2

8.2.1. *Exercise 2.1*

A person has five coins: two of these coins have heads on both sides, one has tails on both sides and the other two are normal coins with heads on one side and tails on the other. It is assumed that all the coins are balanced.

1) The person closes their eyes, chooses a coin at random and tosses it. They obtain heads. We wish to find the probability that the other side will also be heads.

In order to do this, we first model the experience by introducing events. We introduce the following complete system of events.

– A: the coin has heads on both sides;

– B: the coin has tails on both sides;

– C: the coin is normal,

which will enable us to specify the type of coin that was selected. According to this statement, we directly obtain

$$\mathbb{P}(A) = \mathbb{P}(C) = \frac{2}{5}, \mathbb{P}(B) = \frac{1}{5}.$$

We also define the following events

– F_n: the coin turns up heads on the nth toss;

– AF_n: the other side turns up heads on the nth toss.

It is now possible to express the desired probability as $\mathbb{P}(AF_1|F_1)$. We have

$$\mathbb{P}(AF_1|F_1) = \mathbb{P}(AF_1 \cap F_1)\mathbb{P}(F_1).$$

Using the formula for total probabilities, we have

$$\begin{aligned}\mathbb{P}(F_1) &= \mathbb{P}(F_1|A)\mathbb{P}(A) + \mathbb{P}(F_1|B)\mathbb{P}(B) + \mathbb{P}(F_1|C)\mathbb{P}(C)\\ &= 1 \times \frac{2}{5} + 0 \times \frac{1}{5} + \frac{1}{2} \times \frac{2}{5}\\ &= \frac{3}{5},\end{aligned}$$

and again, using similar reasoning, we have

$$\begin{aligned}
&\mathbb{P}(F_1 \cap AF_1) \\
&= \mathbb{P}(F_1 \cap AF_1|A)\mathbb{P}(A) + \mathbb{P}(F_1 \cap AF_1|B)\mathbb{P}(B) + \mathbb{P}(F_1 \cap AF_1|C)\mathbb{P}(C) \\
&= 1 \times \frac{2}{5} + 0 \times \frac{1}{5} + 0 \times \frac{2}{5} \\
&= \frac{2}{5}.
\end{aligned}$$

It follows from this that

$$\mathbb{P}(AF_1|F_1) = \frac{\frac{2}{5}}{\frac{3}{5}} = \frac{2}{3}.$$

2) The person closes their eyes again and tosses the same coin. We wish to find the probability of obtaining heads. We must, therefore, calculate $\mathbb{P}(F_2|F_1)$.

We have $\mathbb{P}(F_2|F_1) = \mathbb{P}(F_2 \cap F_1)\mathbb{P}(F_1)$. Since

$$\begin{aligned}
&\mathbb{P}(F_2 \cap F_1) \\
&= \mathbb{P}(F_2 \cap F_1|A)\mathbb{P}(A) + \mathbb{P}(F_2 \cap F_1|B)\mathbb{P}(B) + \mathbb{P}(F_2 \cap F_1|C)\mathbb{P}(C) \\
&= 1 \times \frac{2}{5} + 0 \times \frac{1}{5} + \frac{1}{4} \times \frac{2}{5} \\
&= \frac{1}{2},
\end{aligned}$$

we then have

$$\mathbb{P}(F_2|F_1) = \mathbb{P}(F_2 \cap F_1)\mathbb{P}(F_1) = \frac{\frac{1}{2}}{\frac{3}{5}} = \frac{5}{6}.$$

3) The person opens their eyes and sees that they have obtained heads. We wish to find the probability that the other side of the coin also shows heads.

Let us calculate

$$\mathbb{P}(AF_2|F_1 \cap F_2) = \mathbb{P}(AF_2 \cap F_1 \cap F_2)\mathbb{P}(F_1 \cap F_2).$$

Further, we have

$$\begin{aligned}
&\mathbb{P}(AF_2 \cap F_2 \cap F_1) \\
&= \mathbb{P}(AF_2 \cap F_2 \cap F_1|A)\mathbb{P}(A) + \mathbb{P}(AF_2 \cap F_2 \cap F_1|B)\mathbb{P}(B) \\
&\quad + \mathbb{P}(AF_2 \cap F_2 \cap F_1|C)\mathbb{P}(C)
\end{aligned}$$

$$= 1 \times \frac{2}{5} + 0 \times \frac{1}{5} + 0 \times \frac{2}{5}$$
$$= \frac{2}{5}.$$

Therefore, it follows

$$\mathbb{P}(AF_2|F_1 \cap F_2) = \mathbb{P}(AF_2 \cap F_1 \cap F_2)\mathbb{P}(F_1 \cap F_2) = \frac{\frac{2}{5}}{\frac{1}{2}} = \frac{4}{5}.$$

4) The person then discards this coin and chooses another one at random and tosses it. We wish to find the probability of obtaining heads. The probability of the discarded coin being A is $4/5$, of it being B is 0 and, thus, the probability of it being C is $1/5$. Let Aj, Bj and Cj be the events where the discarded coin is of types A, B and C, respectively. We have

$$\begin{aligned}
&\mathbb{P}(F)\\
&= \mathbb{P}(F|Aj)\mathbb{P}(Aj) + \mathbb{P}(F|Bj)\mathbb{P}(Bj) + \mathbb{P}(F|Cj)\mathbb{P}(Cj)\\
&= \left(1 \times \frac{1}{4} + 0 \times \frac{1}{4} + \frac{1}{2} \times \frac{1}{2}\right)\frac{4}{5} + \left(1 \times \frac{1}{2} + 0 \times 0 + \frac{1}{2} \times \frac{1}{2}\right)0\\
&\quad + \left(1 \times \frac{1}{2} + 0 \times \frac{1}{4} + \frac{1}{2} \times \frac{1}{4}\right)\frac{1}{5}\\
&= \frac{2}{5} + \frac{5}{8}\\
&= \frac{31}{40}.
\end{aligned}$$

8.2.2. *Exercise 2.2*

The DNA is subject to endogenous and exogenous mutations. In order to survive, the cells have a repairing mechanism. However, sometimes the mutation gets fixed and is passed on to the daughter cells. It is assumed that the number of mutations, M, undergone by the DNA follows a Poisson distribution with parameter λ and p denotes the probability of a mutation being fixed. It is further assumed that the mutations are fixed independently of one another.

1) We wish to find the distribution of the number of fixed mutations F, given that $(M = n)$. The random variable F takes integer values, and we have

$$\mathbb{P}(F = k|M = n) = C_n^k p^k (1-p)^{n-k}.$$

We thus recognize the binomial distribution with parameters n and p, since the fixed mutations are independent.

2) Determine the distribution of the couple (M, F). It must first be noted that the number of fixed mutations cannot exceed the total number of mutations. For $0 \leq k \leq n$, we have

$$\begin{aligned}\mathbb{P}(F = k, M = n) &= \mathbb{P}(F = k|M = n)\mathbb{P}(M = n)\\ &= C_n^k p^k (1-p)^{n-k} \exp(-\lambda)\frac{\lambda^n}{n!}.\end{aligned}$$

3) Let us now determine the distribution of F. We have

$$\begin{aligned}\mathbb{P}(F = k) &= \sum_{n=k}^{\infty} \mathbb{P}(F = k, M = n)\\ &= \sum_{n=k}^{\infty} C_n^k p^k (1-p)^{n-k} \exp(-\lambda)\frac{\lambda^n}{n!}\\ &= \exp(-p\lambda)\frac{(p\lambda)^k}{k!} \exp(-(1-p)\lambda) \sum_{n=k}^{\infty} \frac{((1-p)\lambda)^{n-k}}{(n-k)!}\\ &= \exp(-p\lambda)\frac{(p\lambda)^k}{k!} \exp(-(1-p)\lambda) \exp((1-p)\lambda)\\ &= \exp(-p\lambda)\frac{(p\lambda)^k}{k!}.\end{aligned}$$

We recognize the Poisson distribution with parameter $p\lambda$. Its expectation is therefore $p\lambda$, and its variance is $p\lambda$.

4) Let us now identify the distribution of M given that $(F = k)$. For $n \geq k$, we have

$$\begin{aligned}&\mathbb{P}(M = n|F = k)\\ &= \frac{\mathbb{P}(M = n, F = k)}{\mathbb{P}(F = k)}\\ &= C_n^k p^k (1-p)^{n-k} \frac{\exp(-\lambda)\lambda^n}{n!} \frac{k!}{\exp(-p\lambda)(p\lambda)^k}\\ &= \frac{\exp(-p\lambda)(p\lambda)^k}{k!} \frac{\exp(-(1-p)\lambda)((1-p)\lambda)^{n-k}}{(n-k)!} \frac{k!}{\exp(-p\lambda)(p\lambda)^k}\\ &= \frac{\exp(-(1-p)\lambda)((1-p)\lambda)^{n-k}}{(n-k)!}.\end{aligned}$$

5) Let us finally calculate the conditional expectation of M given F. We have $\mathbb{E}[M|F] = \phi(F)$ with

$$\begin{aligned}
\phi(k) &= \mathbb{E}[M|F = k] \\
&= \sum_n n\mathbb{P}(M = n|F = k) \\
&= \sum_{n=k}^{\infty} n\frac{\exp(-(1-p)\lambda)((1-p)\lambda)^{n-k}}{(n-k)!} \\
&= \exp(-(1-p)\lambda)\sum_{j=0}^{\infty}(j+k)((1-p)\lambda)^j/(j)! \\
&= k + \exp(-(1-p)\lambda)\sum_{j=0}^{\infty} j\frac{((1-p)\lambda)^j}{j!} \\
&= k + (1-p)\lambda,
\end{aligned}$$

recognizing the expectation of a Poisson distribution for the sum. Hence

$$\mathbb{E}[M|F] = F + (1-p)\lambda.$$

8.2.3. *Exercise 2.3*

Let X_1 and X_2 be two independent random variables. It is first assumed that X_1 and X_2 have a binomial distribution with the parameters (n_1, p) and (n_2, p), respectively.

1) Determine the distribution of X_1 given $(X_1 + X_2 = n)$. For any $0 \leq j \leq n$, we have

$$\begin{aligned}
\mathbb{P}(X_1 = j|X_1 + X_2 = n) &= \frac{\mathbb{P}(X_1 = j, X_1 + X_2 = n)}{\mathbb{P}(X_1 + X_2 = n)} \\
&= \frac{\mathbb{P}(X_1 = j, X_2 = n - j)}{\sum_{k=0}^{n} \mathbb{P}(X_1 = k, X_2 = n - k)}.
\end{aligned}$$

By independence, we obtain

$$\begin{aligned}
&\mathbb{P}(X_1 = j|X_1 + X_2 = n) \\
&= \frac{\mathbb{P}(X_1 = j)\mathbb{P}(X_2 = n - j)}{\sum_{k=0}^{n} \mathbb{P}(X_1 = k)\mathbb{P}(X_2 = n - k)} \\
&= \frac{C_{n_1}^{j} p^j (1-p)^{n_1 - j} C_{n_2}^{n-j} p^{n-j} (1-p)^{n_2 - n + j}}{\sum_{k=0}^{n} C_{n_1}^{k} p^k (1-p)^{n_1 - k} C_{n_2}^{n-k} p^{n-k} (1-p)^{n_2 - n + k}}
\end{aligned}$$

$$= \frac{C_{n_1}^j C_{n_2}^{n-j} p^n (1-p)^{n_1+n_2-n}}{\sum_{k=0}^n C_{n_1}^k C_{n_2}^{n-k} p^n (1-p)^{n_1+n_2-n}}$$

$$= \frac{C_{n_1}^j C_{n_2}^{n-j}}{\sum_{k=0}^n C_{n_1}^k C_{n_2}^{n-k}}.$$

It must be noted that the denominator is the number of combinations $C_{n_1+n_2}^n$ recognizing a hypergeometric distribution where we draw n balls from n_1 of type 1 and n_2 of type 2. It may also be noted that since X_1 and X_2 are independent, $X_1 + X_2$ follows the binomial distribution $B(n_1 + n_2, p)$, which directly gives the denominators.

2) Calculate the conditional expectation $\mathbb{E}[X_1|X_1 + X_2]$. We have $\mathbb{E}[X_1|X_1 + X_2] = \phi(X_1 + X_2)$ with

$$\phi(n) = \mathbb{E}[X_1|X_1 + X_2 = n] = n \times \frac{n_1}{n_1 + n_2}$$

as the expectation of the hypergeometric distribution, hence

$$\mathbb{E}[X_1|X_1 + X_2] = (X_1 + X_2)n_1/(n_1 + n_2).$$

It is now assumed that X_1 and X_2 have a Poisson distribution with parameters λ_1 and λ_2, respectively.

3) Let us determine the distribution of X_1 knowing $(X_1 + X_2 = n)$. We have

$$\mathbb{P}(X_1 = j|X_1 + X_2 = n) = \frac{\mathbb{P}(X_1 = j, X_1 + X_2 = n)}{\mathbb{P}(X_1 + X_2 = n)}$$

$$= \frac{\mathbb{P}(X_1 = j, X_2 = n - j)}{\sum_{k=0}^n \mathbb{P}(X_1 = k, X_2 = n - k)}$$

$$= \frac{\exp(-\lambda_1)\frac{\lambda_1^j}{j!}\exp(-\lambda_2)\frac{\lambda_2^{n-j}}{(n-j)!}}{\sum_{k=0}^n \exp(-\lambda_1)\frac{\lambda_1^k}{k!}\exp(-\lambda_2)\frac{\lambda_2^{n-k}}{(n-k)!}}$$

$$= \frac{\lambda_1^j \lambda_2^{n-j}}{j!(n-j)!} \frac{1}{\sum_{k=0}^n \frac{\lambda_1^k \lambda_2^{n-k}}{k!(n-k)!}}$$

$$= \frac{C_n^j(\frac{\lambda_1}{\lambda_1+\lambda_2})^j(\frac{\lambda_2}{\lambda_1+\lambda_2})^{n-j}}{\sum_{k=0}^n C_n^k(\frac{\lambda_1}{(\lambda_1+\lambda_2)})^k(\frac{\lambda_2}{\lambda_1+\lambda_2})^{n-k}}$$

$$= C_n^j(\frac{\lambda_1}{\lambda_1 + \lambda_2})^j(\frac{\lambda_2}{\lambda_1 + \lambda_2})^{n-j}.$$

We recognize the binomial distribution with parameters n and $\dfrac{\lambda_1}{\lambda_1 + \lambda_2}$.

4) Let us calculate the conditional expectation $\mathbb{E}[X_1|X_1 + X_2]$. We have $\mathbb{E}[X_1|X_1 + X_2] = \phi(X_1 + X_2)$ with

$$\phi(n) = \mathbb{E}[X_1|X_1 + X_2 = n] = n \times \lambda_1/(\lambda_1 + \lambda_2)$$

as the expectation of the binomial distribution. Hence

$$\mathbb{E}[X_1|X_1 + X_2] = (X_1 + X_2)\frac{\lambda_1}{\lambda_1 + \lambda_2}.$$

8.2.4. *Exercise 2.4*

Let X be an integrable random variable $\mathcal{F}_0 = \{\emptyset, \Omega\}$. Let us show that $\mathbb{E}[X|\mathcal{F}_0] = \mathbb{E}[X]$.

Since $\mathbb{E}[X|\mathcal{F}_0]$ is $\mathcal{F}_0$-measurable, it is a constant c. Further,

$$c = \mathbb{E}\big[\mathbb{E}[X|\mathcal{F}_0]\big] = \mathbb{E}[X]$$

using the properties of conditional expectation, hence the result.

8.2.5. *Exercise 2.5*

Let (X_1, Y) and (X_2, Y) be two couples of integrable random variables.

1) It is first assumed that the two couples are identically distributed and we will show that in this case $\mathbb{E}[X_1|Y] = \mathbb{E}[X_2|Y]$ a.s. In order to do this, we will use the definition of conditional expectation.

The two random variables $\mathbb{E}[X_1|Y]$ and $\mathbb{E}[X_1|Y]$ are indeed $\sigma(Y)$-measurable, by definition.

Further, for any $A \in \sigma(Y)$, there exists a measurable function f_A such that $\mathbb{1}_A = f_A(Y)$. Since the couples are identically distributed, we have

$$\begin{aligned}
\mathbb{E}[\mathbb{E}[X_1|Y]\mathbb{1}_A] &= \mathbb{E}[X_1\mathbb{1}_A] \\
&= \mathbb{E}[X_1 f_A(Y)] \\
&= \int x f_A(y) d\mathbb{P}_{(X_1,Y)}(x, y) \\
&= \int x f_A(y) d\mathbb{P}_{(X_2,Y)}(x, y) \\
&= \mathbb{E}[X_2 f_A(Y)] \\
&= \mathbb{E}[X_2\mathbb{1}_A],
\end{aligned}$$

therefore, we do have equality a.s.

2) It is now assumed that X_1 takes the values -1 and 1 with the probability $1/2$, $Y = X_1$ and $X_2 = -X_1$. This is thus a case where the marginals are identically distributed, but not the couples. We thus have

$$\mathbb{E}[X_1|Y] = \mathbb{E}[X_1|X_1] = X_1 \text{ } et \text{ } \mathbb{E}[X_2|Y] = \mathbb{E}[-X_1|X_1] = -X_1.$$

Therefore, we have $\mathbb{E}[X_1|Y] \neq \mathbb{E}[X_2|Y]$. Thus, equality of these marginal distributions is not enough for the conditional expectations to be equal.

Let $X_1, \dots, X_n$ be the identically distributed, independent random. We write $T = X_1 + \dots + X_n$.

3) Let us calculate $\mathbb{E}[T|X_1]$. We have $T = X_1 + T'$ with T' independent of X_1. Therefore

$$\mathbb{E}[T|X_1] = X_1 + \mathbb{E}[T'] = X_1 + (n-1)\mathbb{E}[X_1],$$

because X_k have the same distribution as X_1.

4) Let us now calculate $\mathbb{E}[X_1|T]$. Caution: we cannot use the right-hand-side linearity of the condition. We thus start with a slightly different expression, so that we may use the linearity correctly. We have

$$\mathbb{E}[X_1 + \dots + X_n|T] = \mathbb{E}[T|T] = T = \mathbb{E}[X_1|T] + \dots + \mathbb{E}[X_n|T].$$

Further, the couples (X_k, T) are all identically distributed, therefore $\mathbb{E}[X_k|T] = \mathbb{E}[X_1|T]$ from what we have seen before. Therefore, it follows that $\mathbb{E}[X_1|T] = T/n$.

Consider the rolling of a red die and a blue die. Let X_i be the outcome of the ith roll of the red die, and let Y_i be the outcome of the ith roll of the blue die.

5) The couples $(X_1, X_1 + X_2)$ and $(Y_1, Y_1 + Y_2)$ are clearly identically distributed as it is the same experiment and only the color changes.

6) We have $\sigma(X_1 + X_2) \neq \sigma(Y_1 + Y_2)$ since the draws of X only concern the red die (let us call this the first coordinate in ω), and the samples of Y only concern the blue die.

7) According to the aforementioned questions,

$$\mathbb{E}[X_1|X_1 + X_2] = \frac{X_2 + X_2}{2} \text{ } and \text{ } \mathbb{E}[X_2|Y_1 + Y_2] = \frac{Y_1 + Y_2}{2},$$

therefore these conditional expectations are not equal, although the couples $(X_1, X_1 + X_2)$ and $(Y_1, Y_1 + Y_2)$ are identically distributed.

8.2.6. *Exercise 2.6*

Let $(\Omega, \mathcal{F}, \mathbb{P})$ be a probability space, let $\mathcal{G}$ be a sub-σ-algebra of $\mathcal{F}$ and let X be a square-integrable random variable. For any $A \in \mathcal{F}$, we write $\mathbb{P}(A|\mathcal{G}) = \mathbb{E}[\mathbb{1}_A|\mathcal{G}]$. We wish to show that for any $a > 0$,

$$\mathbb{P}(|X| \geq a \,|\mathcal{G}) \leq \frac{1}{a^2}\mathbb{E}[X^2|\mathcal{G}].$$

We have $X^2 \geq X^2\mathbb{1}_{\{X^2 \geq a^2\}}$; therefore, $\mathbb{E}[X^2|\mathcal{G}] \geq \mathbb{E}[X^2\mathbb{1}_{\{X^2 \geq a^2\}}|\mathcal{G}]$.

Further, since $X^2\mathbb{1}_{\{X^2 \geq a^2\}} \geq a^2\mathbb{1}_{\{X^2 \geq a^2\}}$, we have

$$\mathbb{E}[X^2\mathbb{1}_{\{X^2 \geq a^2\}}|\mathcal{G}] \geq a^2\mathbb{E}[\mathbb{1}_{\{X^2 \geq a^2\}}|\mathcal{G}]$$

and $\mathbb{E}[\mathbb{1}_{\{X^2 \geq a^2\}}|\mathcal{G}] = \mathbb{P}[|X| \geq a \,|\mathcal{G}]$. From this, it can be derived that

$$\mathbb{E}[X^2|\mathcal{G}] \geq a^2\mathbb{P}(|X| \geq a \,|\mathcal{G}).$$

8.2.7. *Exercise 2.7*

Let X be a random variable uniformly distributed over the set $\{-2, -1, 1, 2\}$ and let $Y = |X|$.

1) Let us calculate $\mathbb{E}[X]$. Since X is a random variable uniformly distributed over the set $\{-2, -1, 1, 2\}$, we have

$$\mathbb{E}[X] = (-2 - 1 + 1 + 2) \times \frac{1}{4} = 0.$$

2) Let us calculate $\mathbb{E}[X|Y]$. Since Y is a discrete r.v. that takes the values 1 and 2, let us first calculate $\mathbb{E}[X|Y = 1]$ and $\mathbb{E}[X|Y = 2]$.

We have

$$\begin{aligned}\mathbb{E}[X|Y = 1]] = &-1 \times \mathbb{P}(X = -1|Y = 1) - 2 \times \mathbb{P}(X = -2|Y = 1)\\ &+ 1 \times \mathbb{P}(X = 1|Y = 1) + 2 \times \mathbb{P}(X = 2|Y = 1).\end{aligned}$$

On the one hand, we have $\mathbb{P}(X = -2|Y = 1) = \mathbb{P}(X = 2|Y = 1) = 0$. On the other hand, because $(X = -1) \Longrightarrow (Y = 1)$ and

$$\mathbb{P}\,(Y = 1) = \mathbb{P}\,(X = 1) + \mathbb{P}\,(X = -1) = \frac{1}{2},$$

we have

$$\mathbb{P}(X=-1|Y=1)=\frac{\mathbb{P}((X=-1)\cap(Y=1))}{\mathbb{P}(Y=1)}=\frac{\mathbb{P}(X=-1)}{\mathbb{P}(Y=1)}=\frac{1}{2}.$$

In the same way, we obtain that $\mathbb{P}(X=-1|Y=1)=\frac{1}{2}$, which makes it possible to have

$$\mathbb{E}[X|Y=1]=(-1+1)\times\frac{1}{2}=0.$$

In the same way, it can be shown that $\mathbb{E}[X|Y=2]=(-2+2)\times\dfrac{1}{2}=0$, and from this, it can be derived that $\mathbb{E}[X|Y]=0=\mathbb{E}[X]$.

3) Let us show that X and Y are not independent. We have

$$\mathbb{P}((X=1)\cap(Y=1))=\mathbb{P}(X=1)=\frac{1}{4}\neq\mathbb{P}(X=1)\times\mathbb{P}(Y=1)=\frac{1}{4}\times\frac{1}{2}.$$

Therefore, X and Y are not independent.

Conclusion: it is known that if two random variables X and Y are independent, then $\mathbb{E}[X|Y]=\mathbb{E}[X]$. The converse is false. This exercise gives us a counterexample of this.

8.2.8. *Exercise 2.8*

Let X and Y be two independent random variables, each with a Bernoulli distribution with parameter $\frac{1}{2}$.

1) Let us determine the distribution of $X+Y$. We can either use the fact that the sum of two independent r.v.s with a Bernoulli distribution with the parameter p follows the binomial distribution with parameter $(2,p)$, or we can directly carry out the calculation. We have $(X+Y)(\Omega)=\{0,1,2\}$ and, because X and Y are independent, we have

$$\mathbb{P}(X+Y=0)=\mathbb{P}((X=0)\cap(Y=0))=\mathbb{P}(X=0)\times\mathbb{P}(Y=0)=\frac{1}{4},$$

$$\mathbb{P}(X+Y=2)=\mathbb{P}((X=1)\cap(Y=1))=\mathbb{P}(X=1)\times\mathbb{P}(Y=1)=\frac{1}{4},$$

$$\begin{aligned}\mathbb{P}(X+Y=1)&=\mathbb{P}((X=0)\cap(Y=1))+\mathbb{P}((X=1)\cap(Y=0))\\&=\mathbb{P}(X=0)\times\mathbb{P}(Y=1)+\mathbb{P}(X=1)\times\mathbb{P}(Y=0)=\frac{1}{2}.\end{aligned}$$

And we do indeed arrive once again at the binomial distribution with parameters $(2, p)$.

2) Let us calculate $\mathbb{E}[X|X+Y]$ and $\mathbb{E}[Y|X+Y]$. We can either use Exercise 2.3, or we can directly carry out the calculation making use of the fact that we have discrete distributions, which gives us

$$\begin{aligned}
\mathbb{E}[X|X+Y=0] &= \mathbb{P}(X=1|X+Y=0) \\
&= \frac{\mathbb{P}\left(\{X=1\}\cap\{X+Y=0\}\right)}{\mathbb{P}\left(\{X+Y=0\}\right)} \\
&= 0, \\
\mathbb{E}[X|X+Y=2] &= \mathbb{P}(X=1|X+Y=2) \\
&= \frac{\mathbb{P}\left(\{X=1\}\cap\{X+Y=2\}\right)}{\mathbb{P}\left(\{X+Y=2\}\right)} \\
&= \frac{\mathbb{P}\left(\{X=1\}\cap\{Y=1\}\right)}{\mathbb{P}\left(\{X+Y=2\}\right)} \\
&= 1, \\
\mathbb{E}[X|X+Y=1] &= \mathbb{P}(X=1|X+Y=1) \\
&= \frac{\mathbb{P}\left(\{X=1\}\cap\{X+Y=1\}\right)}{\mathbb{P}\left(\{X+Y=1\}\right)} \\
&= 0.
\end{aligned}$$

Therefore, $\mathbb{E}[X|X+Y] = \mathbb{1}_{\{X+Y=2\}} = \frac{X+Y}{2}$. By symmetry, we also have $\mathbb{E}[Y|X+Y] = \mathbb{1}_{\{X+Y=2\}}$.

3) The two random variables $\mathbb{E}[X|X+Y]$ and $\mathbb{E}[Y|X+Y]$ are not independent because, for example,

$$\mathbb{P}\left((\mathbb{E}[X|X+Y]=1)\cap(\mathbb{E}[Y|X+Y]=0)\right)=0$$

but

$$\mathbb{P}\left(\mathbb{E}[X|X+Y]=1\right)=\mathbb{P}(X+Y=2)=\frac{1}{4},$$

and

$$\mathbb{P}\left(\mathbb{E}[Y|X+Y]=0\right)=\mathbb{P}(X+Y\in\{0,1\})=\frac{3}{4}.$$

Conclusion: if X and Y are independent and $\mathcal{G}$ is a sub-σ-algebra, then $\mathbb{E}[X|\mathcal{G}]$ and $\mathbb{E}[Y|\mathcal{G}]$ are not necessarily independent.

8.2.9. *Exercise 2.9*

Let X and Y be two positive, random square-integrable variables, such that $\mathbb{E}[X|Y] = Y$ a.s. and $\mathbb{E}[Y|X] = X$ a.s.

1) Let us show that $\mathbb{E}[X^2] = \mathbb{E}[Y^2] = \mathbb{E}[XY]$. We have

$$\mathbb{E}[X^2] = \mathbb{E}\left[X\mathbb{E}[Y|X]\right] = \mathbb{E}\left[\mathbb{E}[XY|X]\right] = \mathbb{E}[XY].$$

Similarly, we obtain that: $\mathbb{E}[Y^2] = \mathbb{E}[XY]$.

2) Let us show that $X = Y$ a.s. We have

$$\mathbb{E}[(X-Y)^2] = \mathbb{E}[X^2] + \mathbb{E}[Y^2] - 2\mathbb{E}[XY].$$

Using the above, we arrive at $\mathbb{E}[(X-Y)^2] = 0$, which makes it possible to derive that $X = Y$ a.s.

8.2.10. *Exercise 2.10*

Consider a random vector (X, Y) taking values in $\mathbb{N}^2 \setminus \{(0,0)\}$ whose joint distribution is defined by:

$$\forall (i,j) \in \mathbb{N}^2 \setminus \{(0,0)\}, \quad \mathbb{P}(X = i, Y = j) = \frac{(i+j-1)!}{i!j!3^i 6^j \ln 2}.$$

1) Let us verify that this is indeed a probability distribution on $\mathbb{N}^2 \setminus \{(0,0)\}$. In order to do this, it is sufficient to calculate

$$\begin{aligned}
\sum_{(i,j)\in\mathbb{N}^2\setminus\{(0,0)\}} \mathbb{P}(X = i, Y = j) &= \sum_{i=0}^{\infty}\sum_{j=0}^{\infty} \mathbb{1}_{(i,j)\neq(0,0)} \frac{(i+j-1)!}{i!j!3^i6^j \ln 2} \\
&= \frac{1}{\ln 2}\left(\sum_{i=1}^{\infty} \frac{1}{i!3^i} \sum_{j=0}^{\infty} \frac{(i+j-1)!}{j!6^j} + \sum_{j=1}^{\infty} \frac{(j-1)!}{j!6^j}\right) \\
&= \frac{1}{\ln 2}\left(\sum_{i=1}^{\infty} \frac{(i-1)!6^i}{i!3^i5^i} \sum_{j=0}^{\infty} C_{i+j-1}^{j} \frac{1}{6^j}\frac{5^i}{6^i} + \sum_{j=1}^{\infty} \frac{1}{j6^j}\right) \\
&= \frac{1}{\ln 2}\left(\sum_{i=1}^{\infty} \frac{2^i}{i5^i} + \sum_{j=1}^{\infty} \frac{1}{j6^j}\right),
\end{aligned}$$

by recognizing the negative binomial distribution ($\sum_{j=0}^{\infty} C_{i+j-1}^{j} \frac{1}{6^j} \frac{5^i}{6^i} = 1$). It thus follows that

$$\sum_{(i,j)\in\mathbb{N}^2\setminus\{(0,0)\}} \mathbb{P}(X=i,Y=j) = \frac{1}{\ln 2}\Big(-\ln(1-\frac{2}{5}) - \ln(1-\frac{1}{6})\Big)$$
$$= 1,$$

recognizing the Taylor expansion of the logarithmic function: for any $x \in [-1,1[$, we have: $\sum_{j=1}^{+\infty} \frac{x^j}{j} = -\ln(1-x)$. Therefore, we do indeed have a probability distribution.

2) Let us calculate the marginal distribution of X. The random variable X takes values in $\mathbb{N}$. Let us first calculate $\mathbb{P}(X=0)$. We have

$$\begin{aligned}
\mathbb{P}(X=0) &= \sum_{j=1}^{+\infty} \mathbb{P}(X=0,Y=j)\\
&= \sum_{j=1}^{+\infty} \frac{(j-1)!}{j!6^j \ln 2}\\
&= \frac{1}{\ln 2}\sum_{j=1}^{+\infty} \frac{1}{j}\left(\frac{1}{6}\right)^j\\
&= -\frac{ln(1-\frac{1}{6})}{\ln 2}\\
&= \frac{\ln 6 - \ln 5}{\ln 2},
\end{aligned}$$

as seen above.

Let us now calculate $\mathbb{P}(X=i)$ for any $i \in \mathbb{N}\setminus\{0\}$.

$$\begin{aligned}
\mathbb{P}(X=i) &= \sum_{j=0}^{+\infty} \mathbb{P}(X=i,Y=j)\\
&= \sum_{j=1}^{+\infty} \frac{(i+j-1)!}{i!j!3^i6^j \ln 2}\\
&= \frac{1}{i!3^i \ln 2}\sum_{j=1}^{+\infty} \frac{(j+(i-1))!}{j!}\left(\frac{1}{6}\right)^j.
\end{aligned}$$

As it is known that for any $x \in [1-, 1]$ and any natural integer k, we have:

$$\sum_{j=0}^{+\infty} \frac{(j+k)!}{j!} x^j = \frac{k!}{(1-x)^{k+1}},$$

from which it can be derived that:

$$\mathbb{P}(X = i) = \frac{1}{i! 3^i \ln 2} \frac{(i-1)!}{(1-\frac{1}{6})^i} = \frac{2^i}{i 5^i \ln 2}.$$

3) Let us calculate the conditional expectation $\mathbb{E}[Y \mid X]$. Since X is a discrete r.v., let us first calculate $\mathbb{E}[Y \mid X = i]$ for any i. In order to do this, let us differentiate between the two cases $i = 0$ and $i > 0$. We have

$$\begin{aligned}
\mathbb{E}[Y \mid X = 0] &= \sum_{j=1}^{+\infty} j \mathbb{P}(Y = j \mid X = 0) \\
&= \sum_{j=1}^{+\infty} j \frac{\mathbb{P}(Y = j, X = 0)}{\mathbb{P}(X = 0)} \\
&= \frac{1}{(\ln 6 - \ln 5)} \sum_{j=1}^{+\infty} \frac{1}{6^j},
\end{aligned}$$

where, in the last equation, we have used the value of $\mathbb{P}(X = 0)$ obtained in the previous question. Since, for any $x \in [-1, 1]$, we have

$$\sum_{j=1}^{+\infty} x^j = \frac{x}{1-x},$$

It follows from this that

$$\mathbb{E}[Y \mid X = 0] = \frac{1}{5(\ln 6 - \ln 5)}.$$

Now let $i > 0$. We have

$$\begin{aligned}
\mathbb{E}[Y \mid X = i] &= \sum_{j=1}^{+\infty} j \mathbb{P}(Y = j \mid X = i) \\
&= \sum_{j=1}^{+\infty} j \frac{\mathbb{P}(Y = j, X = i)}{\mathbb{P}(X = i)} \\
&= \sum_{j=0}^{+\infty} j \frac{(i+j-1)!}{i! j! 3^i 6^j \ln 2} \frac{i 5^i \ln 2}{2^i}
\end{aligned}$$

$$= \frac{5^i}{(i-1)!6^i} \sum_{j=1}^{+\infty} \frac{((j-1)+i)!}{(j-1)!} \left(\frac{1}{6}\right)^j .$$

$$= \frac{5^i}{(i-1)!6^{i+1}} \sum_{j=0}^{+\infty} \frac{(j+i)!}{j!} \left(\frac{1}{6}\right)^j .$$

$$= \frac{5^i}{(i-1)!6^{i+1}} \frac{i!}{(1-\frac{1}{6})^{i+1}},$$

i.e. $\mathbb{E}[Y \mid X = i] = \frac{i}{5}$.

Conclusion: for any $i \in \mathbb{N}$, we have

$$\mathbb{E}[Y \mid X = i] = \frac{1}{5(\ln 6 - \ln 5)} \mathbb{1}_{i=0} + \frac{i}{5} \mathbb{1}_{i>0},$$

and therefore:

$$\mathbb{E}[Y \mid X] = \frac{1}{5(\ln 6 - \ln 5)} \mathbb{1}_{X=0} + \frac{X}{5} \mathbb{1}_{X>0} = \frac{1}{5(\ln 6 - \ln 5)} \mathbb{1}_{X=0} + \frac{X}{5}.$$

4) We now wish to find a relation between $\mathbb{E}[X]$ and $\mathbb{E}[Y]$. Using the above expression for $\mathbb{E}[Y \mid X]$, we obtain

$$\mathbb{E}[Y] = \mathbb{E}\big[\mathbb{E}[Y \mid X]\big]$$

$$= \frac{1}{5(\ln 6 - \ln 5)} \mathbb{P}(X = 0) + \frac{\mathbb{E}[X]}{5}.$$

By once again using the value of $\mathbb{P}(X = 0)$ that we obtained above, we obtain

$$\mathbb{E}[Y] = \frac{1}{5 \ln 2} + \frac{\mathbb{E}[X]}{5}.$$

8.2.11. *Exercise 2.11*

Let $(\Omega, \mathcal{F}, \mathbb{P})$ be a probability space and let $\mathcal{G}$ be a sub-σ-algebra of $\mathcal{F}$. Let X be a random square-integrable variable $(\Omega, \mathcal{F}, \mathbb{P})$. The conditional variance of X given $\mathcal{G}$ is the random variable

$$\mathbb{V}(X|\mathcal{G}) = \mathbb{E}\Big[\big(X - \mathbb{E}[X|\mathcal{G}]\big)^2 \Big| \mathcal{G}\Big].$$

1) Let us show that $\mathbb{V}(X|\mathcal{G}) = \mathbb{E}[X^2|\mathcal{G}] - \left(\mathbb{E}[X|\mathcal{G}]\right)^2$ a.s. Indeed, we have

$$\begin{aligned}\mathbb{V}(X|\mathcal{G}) &= \mathbb{E}\left[\left(X - \mathbb{E}[X|\mathcal{G}]\right)^2\middle|\mathcal{G}\right]\\ &= \mathbb{E}\left[X^2 - 2X\mathbb{E}[X|\mathcal{G}] + \mathbb{E}[X|\mathcal{G}]^2\middle|\mathcal{G}\right]\\ &= \mathbb{E}[X^2|\mathcal{G}] - \left(\mathbb{E}[X|\mathcal{G}]\right)^2 \; a.s.\end{aligned}$$

because

$$\mathbb{E}\big[X\mathbb{E}[X|\mathcal{G}]|\mathcal{G}\big] = \mathbb{E}[X|\mathcal{G}]\mathbb{E}[X|\mathcal{G}] = \mathbb{E}[X|\mathcal{G}]^2.$$

2) Let us show that $\mathbb{V}(X) = \mathbb{E}\big[\mathbb{V}(X|\mathcal{G})\big] + \mathbb{V}\big(\mathbb{E}[X|\mathcal{G}]\big)$. Indeed, we have

$$\begin{aligned}&\mathbb{E}\big[\mathbb{V}(X|\mathcal{G})\big] + \mathbb{V}\big(\mathbb{E}[X|\mathcal{G}]\big)\\ &= \mathbb{E}\big[\mathbb{E}[X^2|\mathcal{G}] - (\mathbb{E}[X|\mathcal{G}])^2\big] + \mathbb{E}\big[\mathbb{E}[X|\mathcal{G}]^2\big] - \left(\mathbb{E}[\mathbb{E}[X|\mathcal{G}]\right)^2\\ &= \mathbb{E}[X]^2 - (\mathbb{E}[X])^2\\ &= \mathbb{V}(X).\end{aligned}$$

3) Let us now show that $\mathbb{V}(\mathbb{E}[X|\mathcal{G})] \leq \mathbb{V}(X)$. From the definition of conditional variance, we have $\mathbb{V}(\mathbb{E}[X|\mathcal{G}]) \geq 0$. Therefore

$$\mathbb{E}[\mathbb{V}(\mathbb{E}[X|\mathcal{G}])] \geq 0.$$

Using question 2, we arrive at the desired result.

4) Let us now show that $\mathbb{V}(\mathbb{E}[X|\mathcal{G}]) = \mathbb{V}(X)$ if and only if X is $\mathcal{G}$-measurable. In order to do this, we proceed through double implication. First of all, if X is $\mathcal{G}$-measurable, then $\mathbb{E}[X|\mathcal{G}] = X$, and we do indeed have equal variances.

Conversely, let us assume that we have the equality $\mathbb{V}(\mathbb{E}[X|\mathcal{G}]) = \mathbb{V}(X)$. Then, as per question 2, we will have

$$\mathbb{E}[\mathbb{V}(X|\mathcal{G})] = 0 = \mathbb{E}[\mathbb{E}[X - \mathbb{E}[X|\mathcal{G}]]^2|\mathcal{G}] = \mathbb{E}[(X - \mathbb{E}[X|\mathcal{G}])^2];$$

i.e. $X = \mathbb{E}[X|\mathcal{G}]$ a.s. and therefore X is $\mathcal{G}$-measurable.

8.2.12. *Exercise 2.12*

Let X be an integrable and positive random variable on $(\Omega, \mathcal{F}, \mathbb{P})$, and let $\mathcal{G}$ be a sub-σ-algebra of $\mathcal{F}$.

1) Let us show that the event $F = (\mathbb{E}[X|\mathcal{G}] > 0) = \{\omega \in \Omega; \mathbb{E}[X|\mathcal{G}](\omega) > 0\}$ belongs to $\mathcal{G}$.

The function $w \longmapsto \mathbb{E}[X|\mathcal{G}](w)$ is $\mathcal{G}$-measurable, $]0, +\infty[$ is a Borel set and

$$F = \Big(\mathbb{E}[X|\mathcal{G}]\Big)^{-1}(]0, +\infty[).$$

Thus, we do have $F \in \mathcal{G}$.

2) Let us show that $(X > 0) \subset F$.

From the definition of conditional expectation, we have

$$\mathbb{E}\Big[X\mathbb{1}_{\{\mathbb{E}(X|\mathcal{G})=0\}}\Big] = \mathbb{E}\Big[\mathbb{E}[X|\mathcal{G}]\mathbb{1}_{\{\mathbb{E}[X|\mathcal{G}]=0\}}\Big] = 0.$$

Further, X is positive, therefore $X\mathbb{1}_{\{\mathbb{E}[X|\mathcal{G}]=0\}} \geq 0$ a.s. Thus, it follows that $X\mathbb{1}_{\{\mathbb{E}(X|\mathcal{G})=0\}} = 0$ a.s., which implies that

$$\{w \in \Omega; X(w) > 0\} \subset F.$$

3) Let us show that F is the smallest set containing $(X > 0)$ and belonging to $\mathcal{G}$.

Let A be a set belonging to $\mathcal{G}$ and containing $(X > 0)$. In particular, we have $A^c \subset (X = 0)$. Thus, from the definition of conditional expectation, we have

$$0 = \mathbb{E}[X\mathbb{1}_{A^c}] = \mathbb{E}\big[\mathbb{E}[X|\mathcal{G}]\mathbb{1}_{A^c}\big].$$

Furthermore, since X is positive, we have $\mathbb{E}[X|\mathcal{G}] \geq 0$ everywhere, therefore on A^c. Consequently, $\mathbb{E}[X|\mathcal{G}] = 0$ a.s. on A^c. In other words, we do have $F \subset A$ up to a negligible set.

8.3. Solutions to exercises in Chapter 3

8.3.1. *Exercise 3.1*

Let (S_n) be a random walk starting from -1. We wish to find the distribution of the first terms: S_1, S_2, S_3. It is known that

$$S_1 = S_0 + X_1 = -1 + X_1 = \begin{cases} -1 - 1 = -2 \\ -1 + 1 = 0 \end{cases}$$

with the respective probability $1/2$. We then have $S_2 = S_1 + X_2$; therefore, its possible values are

– -3, with a probability of $1/4 = \mathbb{P}(X_1 = X_2 = -1)$;

– -1, with a probability of $1/2 = \mathbb{P}\big((X_1 = 1, X_2 = -1) \cup (X_1 = -1, X_2 = 1)\big)$;

– 1, with a probability of $1/4 = \mathbb{P}(X_1 = X_2 = 1)$.

Using similar reasoning, we obtain that $S_3 = S_2 + X_3$ takes the values

– -4, with a probability of $1/8 = \mathbb{P}(X_1 = X_2 = X_3 = -1)$;

– -2, with a probability of $3/8 = \mathbb{P}\big((X_1 = 1, X_2 = -1, X_3 = -1) \cup (X_1 = -1, X_2 = 1, X_3 = -1) \cup (X_1 = -1, X_2 = -1, X_3 = 1)\big)$;

– 0, with a probability of $3/8 = \mathbb{P}\big((X_1 = -1, X_2 = 1, X_3 = 1) \cup (X_1 = 1, X_2 = -1, X_3 = 1) \cup (X_1 = 1, X_2 = 1, X_3 = -1)\big)$;

– 2, with a probability of $1/8 = \mathbb{P}(X_1 = X_2 = X_3 = 1)$.

8.3.2. *Exercise 3.2*

All possible trajectories of the random walk of four steps, starting from 2 are represented in Figure 8.1.

1) There are $16 = 2^4$ different trajectories, because each step offers two possibilities: climbing or descending.

2) There are $4 = C_4^{2+1}$ paths from $(0, 2)$ to $(4, 0)$.

3) There are $6 = C_4^{2+0}$ paths from $(0, 2)$ to $(4, 2)$, 1 path alone terminates at 2 passing through 0, and 1 path terminates at -2, which is the image of the previous path through the reflection principle.

4) No path ends in 3 or an odd number because we start from an initial even value of $S_0 = 2$ and because there are an even number of steps. We thus have that S_4 is necessarily even.

8.3.3. *Exercise 3.3*

1) All possible paths from $(0, 0)$ to $(3, -1)$ are represented in Figure 8.2. There are $3 = C_3^{\frac{3+1}{2}}$ of these.

2) There is no path from $(0, 0)$ to $(3, 0)$ because it starts from 0, and at the end of three steps, the walk must necessarily take an odd value.

3) There is no path from $(0, 0)$ to $(3, 5)$ because it starts from 0, and at the end of three steps, the walk cannot exceed 3.

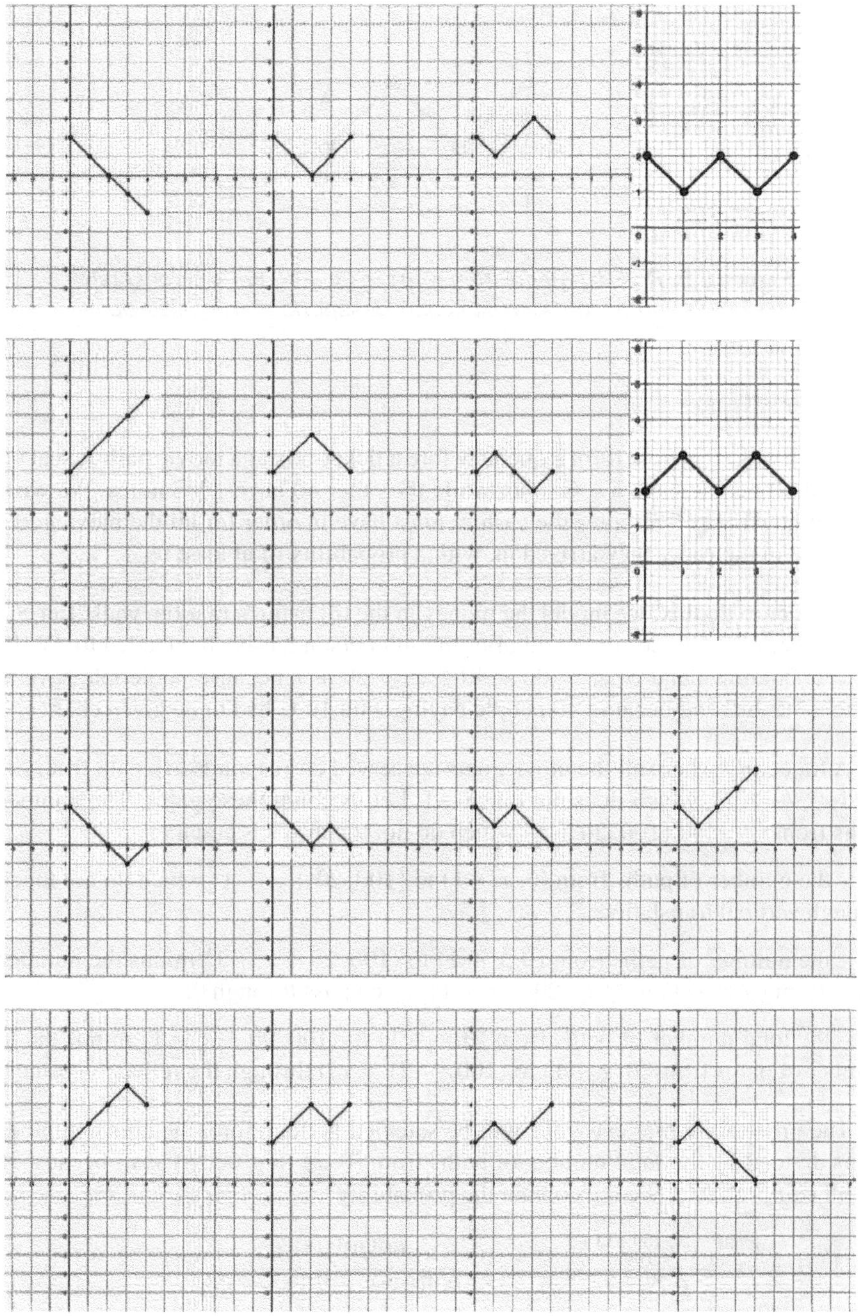

Figure 8.1. *Possible trajectories for the random walk of four steps starting from* 2. *For a color version of this figure, see www.iste.co.uk/desaporta/martingales.zip*

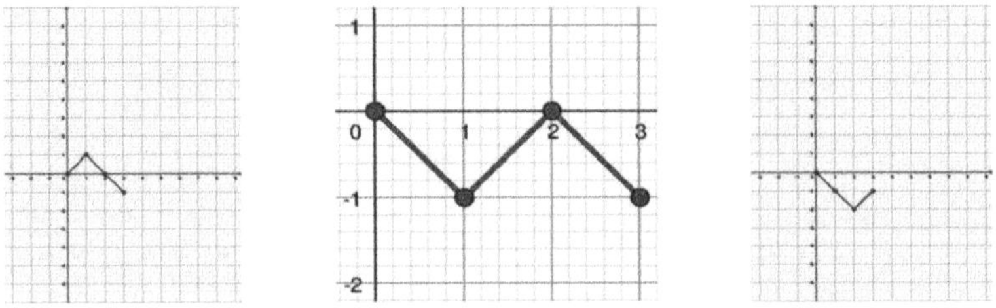

Figure 8.2. *Possible paths from* $(0, 0)$ *to* $(3, -1)$. *For a color version of this figure, see www.iste.co.uk/desaporta/martingales.zip*

8.3.4. *Exercise 3.4*

One hundred people form a queue before a ticket-box at a movie hall. A seat costs 5 € and 60 people have a 5 € note, while the other 40 have 10 € notes. We wish to find out how many 5 € notes the cashier must have in order for all the moviegoers to be served in the order they arrived in, with a probability of at least 95%.

In order to do this, we model the money in the till using a random walk. Let S_n be the number of 5 € notes in the till after the nth person has been attended to. We have $S_0 = x$ unknown and $S_{100} = 60 - 40 + x = 20 + x$ because 60 people have paid using a 5 € bill and because 40 people paying with 10 € had to be given change.

A trajectory where all the moviegoers are served corresponds to a path from $(0, x)$ to $(100, 20 + x)$, which does not touch -1. Let us count these paths. The number of paths from $(0, x)$ to $(100, 20 + x)$ which do not touch -1 is equal to

– the number of paths from $(0, x + 1)$ to $(100, 20 + x + 1)$ which do not touch 0, through vertical translation;

– the number of paths from $(0, x + 1)$ to $(100, 20 + x + 1)$ minus the number of paths from $(0, x + 1)$ to $(100, 20 + x + 1)$ which pass through 0;

– the total number of paths from $(0, x + 1)$ to $(100, 20 + x + 1)$ minus the total number of paths from $(0, x + 1)$ to $(100, -(21 + x))$ through the reflection principle.

The probability that everyone will be satisfied is, therefore, the above number of paths divided by the total number of paths from $(0, x)$ to $(100, 20 + x)$ because they are all equiprobable. We thus obtain the probability

$$
\begin{aligned}
p &= \frac{C_{100}^{60} - C_{100}^{50+11+x}}{C_{100}^{60}} = 1 - \frac{60!40!100!}{100!(61 + x)!(39 - x)!} \\
&= 1 - \frac{60!}{(61 + x)!} \frac{40!}{(39 - x)!}.
\end{aligned}
$$

x	0	1	2	3	4	5
p	0.3443	0.5875	0.7512	0.8562	0.9203	0.9578

Table 8.4. *Values for the probability p of satisfying all moviegoers based on the initial number x of 5 € notes in the till.*

Table 8.4 gives the values of p for x between 0 and 5. It can be seen that from $x = 5$ onwards, this probability exceeds 95%.

8.3.5. *Exercise 3.5*

Consider a non-symmetric random walk: $S_0 = 0$ and $S_n = \sum_{k=1}^{n} X_k$ with $\mathbb{P}(X_k = 1) = 3/4$ and $\mathbb{P}(X_k = -1) = 1/4$.

1) As with a symmetric walk, S_n can take values in the interval $[|-n, n|]$ respecting the fact that S_n has the same parity as n.

For any $n \geq 1$, we write $Y_n = \frac{X_n+1}{2}$.

2) The random variable $Y_1 = (X_n + 1)/2$ takes the values 0 for the event $(X_n = -1)$ whose probability is $1/4$ and 1 for the event $(X_n = 1)$ whose probability is $3/4$. Therefore, it follows a Bernoulli distribution with parameter $3/4$.

3) Let us calculate the distribution of S_n and calculate its expectation and variance. By replacing X_k with $2Y_k - 1$, we obtain

$$S_n = \sum_{k=1}^{n} X_k = \sum_{k=1}^{n} (2Y_k - 1) = 2\left(\sum_{k=1}^{n} Y_k\right) - n.$$

Since X_k are independent, Y_k are also independent. Y_k are independent random variables with a Bernoulli distribution of the same parameter $3/4$; therefore, their sum $Z_n = \sum_{k=1}^{n} Y_k$ follows the binomial distribution with parameter $(n, 3/4)$. Thus, the distribution of S_n is identical to that of $2Z_n - n$. To be more precise, we have

$$\begin{aligned}\mathbb{P}(S_n = k) &= \mathbb{P}(2Z_n - n = k) = \mathbb{P}\big(Z_n = \frac{k+n}{2}\big)\\ &= \begin{cases} C_n^{\frac{k+n}{2}} \left(\frac{1}{4}\right)^{\frac{n-k}{2}} \left(\frac{3}{4}\right)^{\frac{k+n}{2}} & \text{if} \quad k \in [|-n, n|] \text{ and } k+n \text{ even,} \\ 0 & \text{if not.} \end{cases}\end{aligned}$$

In particular, we have

$$\mathbb{E}[S_n] = \mathbb{E}[2Z_n - n] = 2\frac{3n}{4} - n = \frac{n}{2},$$

$$\text{Var}(S_n) = \text{Var}(2Z_n - n) = 4\text{Var}(Z_n) = 4\frac{3n}{4}\frac{1}{4} = \frac{3n}{4}.$$

4) Let us now find the limits for S_n/n and S_n. Since X_k are independent, with the same distribution and integrable, it is possible to apply the strong law of large numbers to them, which give almost sure convergence of $\frac{1}{n}\sum_{k=1}^{n} X_k = \frac{S_n}{n}$ to $\mathbb{E}[X_1] = \frac{1}{2}$. Since this limit is strictly positive, it can be derived from this that $S_n \sim n/2$, therefore S_n tends to $+\infty$ almost surely when n tends to infinity.

8.3.6. *Exercise 3.6*

1) The number of paths from $(0, 0)$ to $(n, 9)$ is equal to

– the number of paths from $(0, -6)$ to $(n, 3)$ through vertical translation;

– the number of paths from $(0, -6)$ to $(n, -3)$ passing through 0, through the reflection principle;

– the number of paths from $(0, 0)$ to $(n, 3)$ passing through 6 by vertical translation.

2) The number of paths from $(0, 0)$ to $(n, a + b)$ is equal to

– the number of paths from $(0, -a)$ to (n, b) through vertical translation;

– the number of paths from $(0, -a)$ to $(n, -b)$ passing through 0 through the reflection principle, because a and b have the same sign;

– the number of paths from $(0, 0)$ to $(n, a - b)$ passing through a through vertical translation.

8.3.7. *Exercise 3.7*

A gambler has 10 € to spend on a slot machine. In each game, they insert 1 € into the machine, which returns 2 € or nothing, with equal probability.

1) Let us model the gambler's fortune using a random walk. Let S_n be the gambler's fortune after the nth game. It is known that $S_0 = 10$, and for any $n \geq 0$, $S_{n+1} = S_n + X_{n+1}$, where X_n represents the winnings of the $n$the game. The random variables X_k are independent, with the same distribution, given by

$$\mathbb{P}(X_1 = -1 + 2) = \mathbb{P}(X_1 = -1 + 0) = \frac{1}{2}.$$

Therefore, $(S_n)_{n\geq 0}$ is indeed a simple symmetric random walk.

Let N be the number of games played until the gambler is ruined.

2) Since the initial wealth is an even figure, the gambler can only be ruined after an even number of games. Further, as they lose a maximum of 1 on each game, they cannot be ruined before the end of the 10th game.

3) Let us calculate the distribution of N. For any $n \geq 5$, we have

$$\begin{aligned}
&\mathbb{P}(N = 2n) \\
&= \mathbb{P}(S_0 = 10, S_1 > 0, \dots, S_{2n-3} > 0, S_{2n-2} > 0, S_{2n-1} > 0, S_{2n} = 0) \\
&= \mathbb{P}(S_0 = 10, S_1 > 0, \dots, S_{2n-3} > 0, S_{2n-2} = 2, S_{2n-1} = 1, S_{2n} = 0).
\end{aligned}$$

The required probability is, therefore, equal to the number of paths from $(0, 10)$ to $(2n, 0)$ that do not touch 0 before $2n$, divided by 2^{2n+1}, which corresponds to the total number of paths of length $2n + 1$.

The number of paths from $(0, 10)$ to $(2n, 0)$ that do not touch 0 before $2n$ is equal to

– the number of paths from $(0, 10)$ to $(2n - 1, 1)$ not touching 0;

– the total number of paths from $(0, 10)$ to $(2n - 1, 1)$ minus the number of paths from $(0, 10)$ to $(2n - 1, 1)$ touching 0;

– the total number of paths from $(0, 10)$ to $(2n - 1, 1)$ minus the total number of paths from $(0, 10)$ to $(2n - 1, -1)$ through the reflection principle.

We thus have

$$\begin{aligned}
\mathbb{P}(N = 2n) &= \frac{1}{2^{2n+1}}(C_{2n-1}^{n-5} - C_{2n-1}^{n-6}) \\
&= \frac{1}{2^{2n+1}} \frac{(2n-1)!}{(n-6)!(n+4)!}(\frac{1}{n-5} - \frac{1}{n+5}) \\
&= \frac{10}{2^{2n+1}} \frac{(2n-1)!}{(n-5)!(n+5)!}.
\end{aligned}$$

8.3.8. *Exercise 3.8*

Two gamblers, Julie and Thomas, play against each other in a game of “Heads or Tails”, using a balanced coin. Before each toss, the two gamblers place 1 € each on the table: if the coin turns up tails, Thomas pockets the 2 €, if it turns up heads,

Julie takes the winnings. At the beginning of the game, Thomas has ten 1 € coins. He does not know Julie's fortune, x. The game stops as soon as one of the two players is ruined.

1) Let us model Thomas' wealth and its evolution using a random walk. Let S_n be Thomas' wealth after the nth toss. It is known that $S_0 = 10$, and for any $n \geq 0$, $S_{n+1} = S_n + X_{n+1}$ where X_n represents the winnings of the nth toss. The random variablesX_k are independent, with the same distribution, given by

$$\mathbb{P}(X_1 = -1) = \mathbb{P}(X_1 = 1) = \frac{1}{2}.$$

Therefore, $(S_n)_{n \geq 0}$ is a simple symmetric random walk.

2) Let us model Julie's wealth and its evolution using a random walk. When Thomas loses, Julie wins and vice versa. Therefore, Julie's winnings after the nth game are R_n with $R_0 = x$ and $R_{n+1} = R_n - X_{n+1}$. Since $-X_k$ has the same distribution as X_k, this is again a simple symmetric random walk.

3) At the initial instant, $S_0 + R_0 = 10 + x$. When Thomas loses, Julie wins and vice versa; therefore, for any integer n, we again have $R_n + S_n = 10 + x$, since there is neither external inflow of money nor any external loss of money.

It is now assumed that Julie's wealth is infinite and that the game stops when Thomas is ruined after 26 tosses..

4) A game that ends with Thomas' ruin after 26 tosses corresponds to a path from $(0, 10)$ to $(26, 0)$, which does not touch 0 before 26.

5) The total number of paths from $(0, 10)$ to $(26, 0)$ is $C_{26}^{13-5} = C_{26}^{8}$.

6) The total number of paths from $(0, 10)$ to $(26, 0)$, which do not touch 0 before 26 is equal to

– the number of paths from $(0, 10)$ to $(25, 1)$ which do not touch 0;

– the total number of paths from $(0, 10)$ to $(25, 1)$ minus the number of paths from $(0, 10)$ to $(25, 1)$ which touch 0;

– the total number of paths from $(0, 10)$ to $(25, 1)$ minus the total number of paths from $(0, 10)$ to $(25, -1)$ through the reflection principle.

This number is thus $C_{25}^{8} - C_{25}^{7}$.

8.4. Solutions to exercises in Chapter 4

8.4.1. *Exercise 4.1*

Let $X \in L^1(\Omega)$. For $n \in \mathbb{N}$, consider the random variable $X_n = \mathbb{E}[X|\mathcal{F}_n]$. We wish to show that the sequence $(X_n)_{n\in\mathbb{N}}$ is an $(\mathcal{F}_n)_{n\in\mathbb{N}}$-martingale. In order to do this, the different points of the definition must be satisfied.

We have $X \in L^1(\Omega)$ by definition. Next, for any $n \in \mathbb{N}$, we have

$$\mathbb{E}[|X_n|] = \mathbb{E}\left[|\mathbb{E}[X|\mathcal{F}_n]|\right] \leq \mathbb{E}\left[\mathbb{E}[|X||\mathcal{F}_n]\right] = \mathbb{E}[|X|] < \infty,$$

where we used the Jensen conditional inequality for the absolute value convex function. Therefore, X_n is integrable for any $n \in \mathbb{N}$.

We also have

$$\mathbb{E}[X_{n+1}|\mathcal{F}_n] = \mathbb{E}\left[\mathbb{E}[X|\mathcal{F}_{n+1}]|\mathcal{F}_n\right] = \mathbb{E}[X|\mathcal{F}_n] = X_n,$$

using the property of conditional expectation for nested σ-algebras. Therefore, (X_n) is indeed an $(\mathcal{F}_n)$-martingale.

8.4.2. *Exercise 4.2*

Let $(X_n)_{n\geq 1}$ be a sequence of positive, independent random variables whose expectation is 1. We write

$$M_0 = 1 \text{ and } M_n = \prod_{i=1}^{n} X_i, \text{ for any } n \geq 1.$$

1) Let us show that $(M_n)_{n\geq 1}$ is a martingale with respect to the natural filtration $(\mathcal{F}_n)_{n\geq 1}$ of the sequence (X_n).

For any integer $n \geq 1$, because X_i are independent, we have

$$\mathbb{E}[M_n] = \mathbb{E}[\prod_{i=1}^{n} X_i] = \prod_{i=1}^{n} \mathbb{E}[X_i] = 1.$$

In addition to this, as M_n is positive, it can be concluded from this that M_n is integrable.

We also have

$$\begin{aligned}\mathbb{E}[M_{n+1}|\mathcal{F}_n] &= \mathbb{E}[M_n X_{n+1}|\mathcal{F}_n]\\ &= M_n\mathbb{E}[X_{n+1}|\mathcal{F}_n]\\ &= M_n\mathbb{E}[X_{n+1}]\\ &= M_n,\end{aligned}$$

since M_n is $\mathcal{F}_n$-measurable by construction and X_{n+1} is independent of $\mathcal{F}_n$. Thus, (M_n) is indeed a martingale.

For $k \geq 1$, we write $a_k = \mathbb{E}[\sqrt{X_k}]$, $N_0 = 1$ and $N_n = \prod_{k=1}^{n} \frac{\sqrt{X_k}}{a_k}$ for $n \geq 1$.

2) Let us show that for any $k \geq 1$, we have $a_k \in]0, 1]$.

On the one hand, $a_k \geq 0$ as it is the expectation of a positive r.v. We even have $a_k > 0$ because otherwise, we would have $X_k = 0$ a.s., which contradicts the fact that the expectation of X_k is equal to 1.

On the other hand, the Cauchy–Schwarz inequality gives $a_k \leq (\mathbb{E}[X_k]^{1/2} = 1$.

3) Let us show that the sequence $(N_n)_{n\geq 0}$ is a martingale with respect to $(\mathcal{F}_n)_{n\geq 0}$.

It is sufficient to apply question 1 to the sequence $(\sqrt{X_n}/a_n)_{n\geq 0}$, which is indeed a sequence of positive, independent random variables and with expectation 1.

8.4.3. *Exercise 4.3*

Let $(X_n)_{n\geq 0}$ be a sequence of integrable random variables and adapted to the filtration $(\mathcal{F}_n)_{n\in\mathbb{N}}$. It is assumed that $\mathbb{E}[X_\tau] = \mathbb{E}[X_0]$ for any bounded stopping time τ.

1) Let $n \geq 0$, $A \in \mathcal{F}_n$ and $\tau = n\mathbb{1}_{A^c} + (n+1)\mathbb{1}_A$. Let us show that τ is an $(\mathcal{F}_n)_{n\geq 0}$-stopping time.

For any $p \in \mathbb{N}$ we have:

$$(\tau = p) = \begin{cases} A^c \; if \; p = n \\ A \; if \; p = n+1 \\ \emptyset \; if \; p \notin \{n, n+1\}. \end{cases}$$

Since A, A^c and $\emptyset$ are all in $\mathcal{F}_n$, we have $(\tau = p) \in \mathcal{F}_n$ for any integer p, and consequently, τ is indeed an $(\mathcal{F}_n)$-stopping time.

2) Let us show that for any $A \in \mathcal{F}_n$ we have

$$\mathbb{E}[X_n \mathbb{1}_{A^c} + X_{n+1} \mathbb{1}_A] = \mathbb{E}[X_n].$$

If we consider the constant stopping time $\tau_1 = n$, then

$$\mathbb{E}[X_{\tau_1}] = \mathbb{E}[X_n] = \mathbb{E}[X_0]. \quad [8.1]$$

If we also consider $\tau_2 = \tau$, then τ_2 is a bounded stopping time, and it follows that

$$\mathbb{E}[X_{\tau_2}] = \mathbb{E}[X_n \mathbb{1}_{A^c} + X_{n+1} \mathbb{1}_A] = \mathbb{E}[X_0]. \quad [8.2]$$

The two equalities above, [8.1] and [8.2], make it possible to obtain the desired result.

3) Let us show that (X_n) is an $(\mathcal{F}_n)_{n \geq 0}$-martingale.

According to the hypothesis, the sequence is integrable and adapted. According to Question 2, for any integer n, for any $A \in \mathcal{F}_n$, we have

$$\mathbb{E}[X_n \mathbb{1}_{A^c} + X_{n+1} \mathbb{1}_A] = \mathbb{E}[X_n].$$

Thus, for any integer n and for any $A \in \mathcal{F}_n$ we have

$$\begin{aligned} \mathbb{E}[X_n \mathbb{1}_A] &= \mathbb{E}[X_n] - \mathbb{E}[X_n \mathbb{1}_{A^c}] \\ &= \mathbb{E}[X_n \mathbb{1}_{A^c} + X_{n+1} \mathbb{1}_A] - \mathbb{E}[X_n \mathbb{1}_{A^c}] \\ &= \mathbb{E}[X_{n+1} \mathbb{1}_A], \end{aligned}$$

which implies that $\mathbb{E}[X_{n+1} | \mathcal{F}_n] = \mathbb{E}[X_n]$. Thus, the sequence is indeed an $(\mathcal{F}_n)$-martingale.

8.4.4. *Exercise 4.4*

Consider $(X_n)_{n \geq 0}$ a sequence of random variables taking values in $\{0, 1\}$, we write:

$$X_0 = 1 \text{ and }, \forall n \in \mathbb{N}, \ S_n = \sum_{p=0}^{n} X_p,$$

and it is assumed that for any $n \in \mathbb{N}$, the distribution of X_{n+1} given $(X_0, ..., X_n)$ is the Bernoulli distribution with parameter $M_n = \dfrac{S_n}{n+2}$. Finally, we consider $\mathcal{F} = (\mathcal{F}_n)_{n \in \mathbb{N}}$ the natural filtration of $(X_n)_{n \in \mathbb{N}}$.

1) Let us calculate $\mathbb{E}[X_{n+1} \mid \mathcal{F}_n]$.

The distribution of X_{n+1} given $(X_0, ..., X_n)$ is the Bernoulli distribution with parameter M_n, thus $\mathbb{P}_{X_{n+1}}(\cdot \mid \mathcal{F}_n) = (1 - M_n)\delta_0 + M_n\delta_1$. Consequently, $\mathbb{E}[X_{n+1} \mid \mathcal{F}_n] = M_n$.

2) Let us now calculate $\mathbb{E}[S_{n+1} \mid \mathcal{F}_n]$.

For any $n \geq 0$, we have

$$\begin{aligned}\mathbb{E}[S_{n+1} \mid \mathcal{F}_n] &= \sum_{p=0}^{n+1} \mathbb{E}[X_p \mid \mathcal{F}_n] = \sum_{p=0}^{n} X_p + \mathbb{E}[X_{n+1} \mid \mathcal{F}_n] \\ &= S_n + \mathbb{E}[X_{n+1} \mid \mathcal{F}_n].\end{aligned}$$

From the above question, it can be derived that

$$\mathbb{E}[S_{n+1} \mid \mathcal{F}_n] = S_n + M_n = S_n + \frac{S_n}{n+2} = \frac{(n+3)S_n}{n+2}.$$

3) Let us show that $(M_n)_{n \in \mathbb{N}}$ is an $\mathcal{F}$-martingale.

For any $n \in \mathbb{N}$, we have $\mid M_n \mid \leq 1$ almost surely; thus, for any $n \in \mathbb{N}$, M_n is integrable. Further, for any $n \in \mathbb{N}$,

$$\mathbb{E}[M_{n+1} \mid \mathcal{F}_n] = \mathbb{E}\Big[\frac{S_{n+1}}{n+3} \mid \mathcal{F}_n\Big] = \frac{1}{n+3}\mathbb{E}\Big[S_{n+1} \mid \mathcal{F}_n\Big] = \frac{S_n}{n+2},$$

according to the previous question. Consequently, for any $n \in \mathbb{N}$,

$$\mathbb{E}[M_{n+1} \mid \mathcal{F}_n] = M_n,$$

which implies that $(M_n)_{n \in \mathbb{N}}$ is indeed an $\mathcal{F}$-martingale.

8.4.5. *Exercise 4.5*

Let (X_n) be an $(\mathcal{F}_n)$-martingale and let T be an $(\mathcal{F}_n)$-stopping time such that $\mathbb{E}[T] < \infty$, and there exists a constant C such that for any integer n,

$$\mathbb{E}[|X_{n+1} - X_n| \mid \mathcal{F}_n] \leq C.$$

1) Let us show that $\mathbb{P}(T < \infty) = 1$.

Let us use absurd reasoning and assume that $\mathbb{P}(T = \infty) > 0$. Since

$$\mathbb{P}(T = \infty) = \mathbb{P}\left(\cap_{n=1}^{+\infty}\{T \geq n\}\right) = \lim_{n\to\infty} \mathbb{P}(T \geq n),$$

we obtain that $\lim_{n\to\infty} \mathbb{P}(T \geq n) > 0$. Further, we have

$$\lim_{n\to\infty} \mathbb{P}\{T \geq n\} \leq \lim_{n\to\infty} \frac{\mathbb{E}[T]}{n} = 0,$$

from the Markov inequality, using the hypothesis $\mathbb{E}[T] < \infty$. We thus have a contradiction. Therefore, we have $\mathbb{P}(T = \infty) = 0$ and consequently, $\mathbb{P}(T < \infty) = 1$.

We write $W = |X_0| + \sum_{0\leq k} \mathbb{1}_{\{k<T\}}|X_{k+1} - X_k|$.

2) Let us show that $\sup_{n\geq 0} |X_{T\wedge n}| \leq W$.

For any $n \in \mathbb{N}$ we have:

$$X_{T\wedge n} = X_0 + \sum_{k=0}^{T\wedge n-1} (X_{k+1} - X_k) = X_0 + \sum_{k\geq 0} \mathbb{1}_{\{k<T\wedge n\}}(X_{k+1} - X_k).$$

Since $\mathbb{1}_{\{k<T\wedge n\}} = \mathbb{1}_{\{k<T\}}1_{\{k<n\}}$ and $\mathbb{1}_{\{k<n\}} = 1$ if $k = 0, 1, \ldots., n-1$, and 0 if $k \geq n$, we obtain that

$$X_{T\wedge n} = X_0 + \sum_{k=0}^{n-1} \mathbb{1}_{\{k<T\}}(X_{k+1} - X_k).$$

Therefore, we have

$$|X_{T\wedge n}| \leq |X_0| + \sum_{k=0}^{n-1} \mathbb{1}_{\{k<T\}}|X_{k+1} - X_k| = W.$$

Consequently, we do indeed have $\sup_{n\geq 0} |X_{T\wedge n}| \leq W$.

3) Let us show that $\mathbb{E}[W] < \infty$.

For any $k \geq 0$, since $\mathbb{1}_{\{k<T\}}$ is $\mathcal{F}_k$-measurable, we have

$$\begin{aligned}\mathbb{E}\left[\mathbb{1}_{\{k<T\}}|X_{k+1} - X_k|\right] &= \mathbb{E}\left[\mathbb{E}\left[\mathbb{1}_{\{k<T\}}|X_{k+1} - X_k|\ |\mathcal{F}_k\right]\right]\\ &= \mathbb{E}\left[\mathbb{1}_{\{k<T\}}\mathbb{E}\left[|X_{k+1} - X_k|\ |\mathcal{F}_k\right]\right].\end{aligned}$$

Also, since according to the hypothesis $\mathbb{E}[|X_{k+1} - X_k||\mathcal{F}_k] \leq C$, it follows that

$$\mathbb{E}\left[\mathbb{1}_{\{k<T\}}|X_{k+1} - X_k|\right] \leq C\mathbb{E}\left[\mathbb{1}_{\{k<T\}}\right] = C\mathbb{P}(T > k).$$

From this, it can be derived that

$$\begin{aligned}
\mathbb{E}[W] &= \mathbb{E}[|X_0|] + \sum_{0\leq k} \mathbb{E}\left[\mathbb{1}_{\{k<T\}}|X_{k+1} - X_k|\right] \\
&\leq \mathbb{E}[|X_0|] + C\sum_{0\leq k} \mathbb{P}(k < T) \\
&= \mathbb{E}[|X_0|] + C\sum_{0\leq k} \mathbb{E}[\mathbb{1}_{\{k<T\}}] \\
&= \mathbb{E}[|X_0|] + C\mathbb{E}[\sum_{0\leq k} \mathbb{1}_{\{k<T\}}] \\
&= \mathbb{E}[|X_0|] + C\mathbb{E}[T] < +\infty.
\end{aligned}$$

4) Let us show that $\mathbb{E}[X_T] = \mathbb{E}[X_0]$.

As the hypotheses in the stopping time Theorem 4.4 are satisfied, we directly obtain that $\mathbb{E}[X_T] = \mathbb{E}[X_0]$.

Let $Z_0 = 0$ be the identically zero r.v., and let $(Z_n)_{n\geq 1}$ be a sequence of independent, centered and i.i.d. random variables. In addition, let T be a stopping time adapted with respect to the natural filtration $(\mathcal{G}_n)_{n\geq 0}$ of the process $(Z_n)_{n\geq 0}$, and such that $\mathbb{E}[T] < \infty$. Consider the sequence of random variables defined by $X_n = \sum_{k=0}^{n} Z_k$, for any $n \geq 0$.

5) Let us show that $\mathbb{E}[X_T] = \mathbb{E}[X_0] = 0$.

As the sequence $(X_n)_{n\geq 0}$ is a martingale with respect to $(\mathcal{G}_n)_{n\geq 0}$, in order to arrive at the required result, it is sufficient to verify that there exists $C > 0$ such that for any integer n, $\mathbb{E}[|X_{n+1} - X_n||\mathcal{G}_n] \leq C$ (according to Question 4.)

Further, $Z_{n+1} = X_{n+1} - X_n$ is, by hypothesis, independent of $\mathcal{G}_n$. It follows from this that for any $n \geq 0$, we have:

$$\mathbb{E}[|X_{n+1} - X_n||\mathcal{G}_n] = \mathbb{E}[|Z_{n+1}|] = \mathbb{E}[|Z_1|].$$

as Z_n have the same distribution. By taking $C = \mathbb{E}[|Z_1|] < \infty$, we arrive at the result.

Now, let $Z_0 = a$ be a constant r.v. and let $(Z_n)_{n\geq 1}$ be a sequence of integrable, i.i.d. random variables that are no longer assumed to be centered. It is also assumed that $\mathbb{E}[Z_1] = a$. Let T still be a stopping time adapted to the natural filtration $(\mathcal{G}_n)_{n\geq 0}$ of the process $(Z_n)_{n\geq 0}$, and such that $\mathbb{E}[T] < \infty$. We again consider a sequence of random variables defined by $X_n = \sum_{k=0}^{n} Z_k$, for any $n \geq 0$.

6) Let us show that $\mathbb{E}[X_T] = a\mathbb{E}[T]$.

We consider the sequence $(Y_n)_{n\geq 1}$ defined by: $Y_0 = 0$ and for any $n \geq 1$, $Y_n = Z_n - a$. We also consider the sequence, $(X'_n)_{n\geq 0}$, defined by

$$X'_n = \sum_{k=0}^{n} Y_n = \sum_{k=0}^{n} Z_n - na = X_n - na.$$

The sequence (X'_n) is a $(\mathcal{G}_n)_{n\geq 0}$-martingale since Y_n are centered and, in addition, we have $X'_T = X_T - aT$. On the one hand, according to Question 5, we have $\mathbb{E}[X'_T] = 0$.

On the other hand, we have $\mathbb{E}[X'_T] = \mathbb{E}[X_T] - a\mathbb{E}[T]$. It can thus be derived from this that $\mathbb{E}[X_T] = a\mathbb{E}[T]$.

7) We also assume that $\mathbb{E}[Z_1^2] < \infty$. Let us show that

$$\mathbb{E}\left[\sum_{k=1}^{T}(Z_k - a)^2\right] = \mathbb{V}(Z_1)\mathbb{E}[T].$$

Consider the r.v. (K_n) defined by: $K_0 = 0$ and for any $n \geq 1$,

$$K_n = \sum_{k=1}^{n}(Z_n - a)^2 - n\mathbb{V}(Z_1).$$

For any $n \geq 1$, K_n is a sequence of n centered and i.i.d. random variables. By applying Question 5 to (K_n), we obtain that $\mathbb{E}\left[\sum_{k=1}^{T}(Z_k - a)^2 - T\mathbb{V}(Z_1)\right] = 0$, which implies that

$$\mathbb{E}\left[\sum_{k=1}^{T}(Z_k - a)^2\right] = \mathbb{V}(Z_1)\mathbb{E}[T].$$

8.4.6. *Exercise 4.6*

Let $p \in \left]\frac{1}{2}, 1\right[$, and let $(X_n)_{n\in\mathbb{N}}$ be the sequence of random variables taking values in $\mathbb{Z}$ defined by: $X_0 = 0$ and for any $n \geq 1$, $X_n = Y_1 + ... + Y_n$, where (Y_n) is a sequence of i.i.d. random variables and

$$\mathbb{P}(Y_n = 1) = p \text{ and } \mathbb{P}(Y_n = -1) = 1 - p.$$

The sequence (X_n) is therefore an asymmetric random walk. Let $(\mathcal{F}_n)_{n\geq 0}$ be the natural filtration of (Y_n).

1) We wish to find the set S of real numbers $\alpha \in \mathbb{R}_+^\star$ such that $\left(\alpha^{X_n}\right)_{n\in\mathbb{N}}$ is an $(\mathcal{F}_n)_{n\geq 0}$-martingale.

Let $\alpha \in \mathbb{R}_+^\star$. Since the sequence (Y_n) is i.i.d., we have

$$\mathbb{E}\left[\left|\alpha^{X_n}\right|\right] = \mathbb{E}\left[\alpha^{X_n}\right] = \mathbb{E}\left[\prod_{i=1}^{n} \alpha^{Y_i}\right] = \prod_{i=1}^{n} \mathbb{E}\left[\alpha^{Y_i}\right].$$

Further, for any $i \in \{1, ..., n\}$,

$$\mathbb{E}\left[\alpha^{Y_i}\right] = \alpha p + \alpha^{-1}(1-p).$$

Therefore,

$$\mathbb{E}\left[\left|\alpha^{X_n}\right|\right] = (\alpha p + \alpha^{-1}(1-p))^n < \infty,$$

and consequently, for any $\alpha \in \mathbb{R}_+^\star$ and any $n \in \mathbb{N}$, α^{X_n} is integrable.

On the other hand, for any $n \in \mathbb{N}$, since the sequence $(\alpha^{X_n})_{n\in\mathbb{N}}$ is $(\mathcal{F}_n)_{n\in\mathbb{N}}$-adapted by construction, we have

$$\mathbb{E}\left[\alpha^{X_{n+1}}|\mathcal{F}_n\right] = \mathbb{E}\left[\alpha^{X_n+Y_{n+1}}|\mathcal{F}_n\right] = \alpha^{X_n}\mathbb{E}\left[\alpha^{Y_{n+1}}|\mathcal{F}_n\right],$$

In addition, because the sequence of random variables (Y_n) is independent, in particular, Y_{n+1} is independent of $\mathcal{F}_n$ and we have

$$\mathbb{E}\left[\alpha^{Y_{n+1}}|\mathcal{F}_n\right] = \mathbb{E}\left[\alpha^{Y_{n+1}}\right] = \alpha p + \alpha^{-1}(1-p).$$

Consequently, we have

$$\mathbb{E}\left[\alpha^{X_{n+1}}|\mathcal{F}_n\right] = \left(\alpha p + \alpha^{-1}(1-p)\right)\alpha^{X_n}.$$

From this, it can be derived that $(\alpha^{X_n})_{n\in\mathbb{N}}$ is a martingale if and only if,

$$\alpha p + \alpha^{-1}(1-p) = 1,$$

which is equivalent to the second-degree equation using α;

$$p\alpha^2 - \alpha + 1 - p = 0,$$

whose solutions are $\alpha = 1$ and $\alpha = \dfrac{1-p}{p}$. From this, it can be derived that the desired set is $S = \left\{1, \dfrac{1-p}{p}\right\}$.

For the rest of this exercise, we will take $\alpha \in S$. Let $a, b \geq 0$ and

$$T = \min\{n \in \mathbb{N}; X_n = -a \text{ ou } X_n = b\}.$$

2) Let us show that T is an $(\mathcal{F}_n)_{n\geq 0}$-stopping time.

Writing $A = \{-a, b\}$, for any $n \in \mathbb{N}^*$ we have:

$$(T = n) = \bigcap_{i=0}^{n-1}(X_i \in \mathbb{R} \setminus A) \cap (X_n \in A) \in \mathcal{F}_n,$$

because A is a Borel set and each of the events $\{X_i \in \mathbb{R} \setminus A\}$ and $\{X_n \in A\}$ belongs to $\mathcal{F}_n$. Further, since,

$$(T = 0) = (X_0 = -a) \cup (X_0 = b) \in \mathcal{F}_0,$$

it can be derived from this that T is an $(\mathcal{F}_n)_{n\geq 0}$-stopping time.

3) Let us show that the sequence of random variables (X_n) has a.s. convergence to $+\infty$.

According to the strong law of large numbers, the sequence $\left(\dfrac{X_n}{n}\right)$ converges a.s. to $\mathbb{E}[Y_1] = 2p - 1 > 0$. Therefore,

$$\mathbb{P}(\lim_{n\to\infty} X_n = \lim_{n\to\infty} (2p-1)n) = 1,$$

and consequently,

$$\mathbb{P}(\lim_{n\to\infty} X_n = +\infty) = 1,$$

which gives the a.s. convergence of (X_n) to $+\infty$.

4) Let us show that $T < \infty$ a.s.

We have $\mathbb{P}(\lim_{n\to\infty} X_n = +\infty) = 1$ and (X_n) is a simple asymmetric walk starting from 0. In particular, the walk can only have steps with amplitude 1. Thus, we have

$$\mathbb{P}(\omega \in \Omega; \exists n_0(\omega) \in \mathbb{N}; X_{n_0(\omega)}(\omega) = b) = 1.$$

because it starts from 0, it must pass through every positive integer to reach infinity. From this, it follows that

$$\mathbb{P}(\omega \in \Omega; T(\omega) < \infty) = 1.$$

5) Let us now show that for any $t \geq 0$, $\mathbb{E}[\alpha^{X_{T\wedge t}}] = 1$.

The stopping time $T \wedge t$ is bounded by t, thus, according to the stopping theorem, we have

$$\mathbb{E}[\alpha^{X_{T\wedge t}}] = \mathbb{E}[\alpha^{X_{T\wedge 0}}] = \mathbb{E}[\alpha^{X_0}] = \mathbb{E}[\alpha^0] = 1.$$

6) Let us show that $\mathbb{E}[\alpha^{X_T}] = 1$.

Since $T < \infty$ a.s., we have $\lim_{t\to+\infty} X_{T\wedge t} = X_T$ a.s.. Further, because $\alpha^{X_{T\wedge t}}$ is bounded by $\max(\alpha^{-a}, \alpha^b)$, using the dominated convergence theorem and the previous question, we obtain:

$$\mathbb{E}[\alpha^{X_T}] = \lim_{t\to+\infty} \mathbb{E}[\alpha^{X_{T\wedge t}}] = 1.$$

7) Let us calculate $\mathbb{P}(X_T = -a)$.

We have

$$1 = \mathbb{E}[\alpha^{X_T}] = \alpha^{-a}\mathbb{P}(X_T = -a) + \alpha^{-b}(1 - \mathbb{P}(X_T = -a)),$$

which implies that

$$\mathbb{P}(X_T = -a) = \frac{\alpha^a - \alpha^{a+b}}{1 - \alpha^{a+b}}.$$

8.4.7. *Exercise 4.7*

Let $(Z_n)_{n\geq 1}$ be a sequence of independent random variables taking values in $\{-1, 0, 1\}$ such that $Z_0 = 0$ a.s. and

$$\forall n \geq 1,\ \mathbb{P}(Z_n = 1) = \mathbb{P}(Z_n = -1) = \frac{1}{2}.$$

We consider $\mathcal{F} = (\mathcal{F}_n)_{n\in\mathbb{N}}$ the natural filtration of $(Z_n)_{n\in\mathbb{N}}$

$$S_n = \sum_{i=0}^{n} Z_i,$$

for any $n \geq 0$. The sequence (S_n) is therefore the simple symmetric walk starting from 0. Let $a \in \mathbb{N}^*$.

1) Let us show that $T_a := \inf\{n \geq 0; |S_n| \geq a\}$ is an $\mathcal{F}$-stopping time.

Writing $A =]-\infty, -a] \cup [a, +\infty[$, it is possible to write $T_a = \inf\{n \geq 0, S_n \in A\}$. Therefore, T_a is the hitting time of the Borel set A

(cf. Example 4.8), and consequently, T_a is a stopping time with respect to the natural filtration of (S_n), which is denoted by $\mathcal{G}$.

Since, in addition, $S_n = \sum_{i=0}^{n} Z_i$ for any $n \in \mathbb{N}$, we have $\mathcal{G} \subset \mathcal{F}$ and consequently, T_a is an $\mathcal{F}$-stopping time.

Let us write $T'_a = \inf\{n \in \mathbb{N}; |S_n| = a\}$.

2) Let us show that $T_a = T'_a$.

For any $\omega \in \Omega$, we have

$$\{n \in \mathbb{N}; |S_n(\omega)| = a\} \subset \{n \in \mathbb{N}; |S_n(\omega)| \leq a\},$$

and consequently,

$$T_a(\omega) \leq T'_a(\omega).$$

– If $T_a(\omega) = +\infty$, then the above inequality implies that $T'_a(\omega) = +\infty$, and consequently, there is equality.

– If $T_a(\omega) < +\infty$, then by definition T_a, we have $T_a(\omega) \geq 1$ and

$$|S_{T_a(\omega)-1}(\omega)| \leq a - 1 \leq |S_{T_a(\omega)}(\omega)| - 1.$$

On the other hand, we have $|S_{T_a(\omega)}(\omega) - S_{T_a(\omega)-1}(\omega)| = 1$, therefore

$$\begin{aligned}
|S_{T_a(\omega)-1}(\omega)| &= |S_{T_a(\omega)-1}(\omega) - S_{T_a(\omega)}(\omega) + S_{T_a(\omega)}(\omega)| \\
&\geq |S_{T_a(\omega)}(\omega)| - |S_{T_a(\omega)-1}(\omega) - S_{T_a(\omega)}(\omega)| \\
&\geq |S_{T_a(\omega)}(\omega)| - 1 \\
&\geq a - 1.
\end{aligned}$$

From this, we can derive that $|S_{T_a(\omega)}(\omega)| = a$, and consequently,

$$T'_a(\omega) \leq T_a(\omega).$$

Finally, we do indeed have $T_a = T'_a$ a.s.

Let us fix $\lambda \in]0, \frac{\pi}{2a}[$ and for any $n \in \mathbb{N}$, we consider the random variable

$$W_n = \cos(\lambda S_n)(\cos(\lambda))^{-n}.$$

3) Let us show that $(W_n)_{n\in\mathbb{N}}$ is an $\mathcal{F}$-martingale.

For any $n \in \mathbb{N}$, we have

$$|W_n| \leq (\cos(\lambda))^{-n}.$$

Therefore, for any $n \in \mathbb{N}$, the random variable W_n is integrable.

We also have

$$\begin{aligned}
&(\cos\lambda)^{n+1}\mathbb{E}\left[W_{n+1}|\mathcal{F}_n\right] \\
&= \mathbb{E}\left[\cos(\lambda S_{n+1})|\mathcal{F}_n\right] \\
&= \mathbb{E}\left[\cos(\lambda S_n + \lambda Z_{n+1})|\mathcal{F}_n\right] \\
&= \mathbb{E}\left[\cos(\lambda S_n)\cos(\lambda Z_{n+1}) - \sin(\lambda S_n)\sin(\lambda Z_{n+1})|\mathcal{F}_n\right] \\
&= \mathbb{E}\left[\cos(\lambda S_n)\cos(\lambda Z_{n+1})|\mathcal{F}_n\right] - \mathbb{E}\left[\sin(\lambda S_n)\sin(\lambda Z_{n+1})|\mathcal{F}_n\right] \\
&= \cos(\lambda S_n)\mathbb{E}\left[(\cos(\lambda Z_{n+1}))\,|\mathcal{F}_n\right] - \sin(\lambda S_n)\mathbb{E}\left[\sin(\lambda Z_{n+1}))|\mathcal{F}_n\right],
\end{aligned}$$

where the last equality arises from the fact that S_n is $\mathcal{F}_n$-measurable. Further, Z_{n+1} is independent of $\mathcal{F}_n$. Therefore, using the distribution of Z_{n+1}, we have

$$\begin{aligned}
\mathbb{E}\left[\cos(\lambda Z_{n+1})|\mathcal{F}_n\right] &= \mathbb{E}\left[\cos(\lambda Z_{n+1})\right] \\
&= \frac{1}{2} \times \cos\lambda + \frac{1}{2} \times \cos(-\lambda) \\
&= \cos\lambda
\end{aligned}$$

and

$$\begin{aligned}
\mathbb{E}\left[\sin(\lambda Z_{n+1})|\mathcal{F}_n\right] &= \mathbb{E}\left[\sin(\lambda Z_{n+1})\right] \\
&= \frac{1}{2} \times \sin\lambda + \frac{1}{2} \times \sin(-\lambda) \\
&= 0.
\end{aligned}$$

From this, it follows that

$$\mathbb{E}\left[W_{n+1}|\mathcal{F}_n\right] = (\cos\lambda)^{-(n)}\cos(\lambda S_n) = W_n.$$

From this, it can be derived that the sequence (W_n) is an $(\mathcal{F}_n)$-martingale.

4) Let us show that for any $n \in \mathbb{N}$, we have

$$\cos(\lambda a)(\cos(\lambda))^{-\min(n,T_a)} \leq W_{\min(n,T_a)}.$$

We have $\min(n, T_a) \leq T_a$. Therefore, for any $\lambda \in]0, \frac{\pi}{2a}[$ we have:

$$0 \leq \lambda|S_{\min(n,T_a)}| \leq \lambda a \leq \frac{\pi}{2}.$$

Since the cosine function decreases on the interval $[0, \frac{\pi}{2a}]$, it can be derived from this that

$$\cos(\lambda a) \leq \cos\left(\lambda |S_{\min(n,T_a)}|\right) = \cos\left(\lambda S_{min(n,T_a(w))}(w)\right),$$

and consequently,

$$(\cos\lambda)^{-\min(n,T_a)}\cos(\lambda a) \leq (\cos\lambda)^{-\min(n,T_a)}\cos\left(\lambda S_{\min(n,T_a)}\right),$$

or again

$$(\cos\lambda)^{-\min(n,T_a)}\cos(\lambda a) \leq W_{\min(n,T_a)}.$$

5) Let us show that for any $n \in \mathbb{N}$,

$$\mathbb{E}\Big[\cos(\lambda a)(\cos(\lambda))^{-\min(n,T_a)}\Big] \leq 1.$$

Using the inequality obtained in Question 4 and the fact that the two terms of the inequality are positive, we obtain that

$$\mathbb{E}\left[(\cos\lambda)^{-\min(n,T_a)}\cos(\lambda a)\right] \leq \mathbb{E}\left[W_{\min(n,T_a)}\right].$$

Further, since T_a is a stopping time and (W_n) is a martingale, according to Proposition 4.5, the sequence $(W_{\min(n,T_a)})$ is also a martingale. Thus, it follows that

$$\mathbb{E}\left[W_{\min(n,T_a)}\right] = \mathbb{E}\left[W_{\min(0,T_a)}\right].$$

As we also have

$$W_{\min(0,T_a)} = W_0 = \cos(\lambda S_0) = \cos 0 = 1 \text{ a.s.},$$

it can be derived from this that

$$\mathbb{E}\left[W_{\min(n,T_a)}\right] \leq 1.$$

8.4.8. *Exercise 4.8*

Let $(X_n)_{n\geq 0}$ be a sequence of random variables defined on the same probability space. This sequence is said to be *uniformly integrable* if we have:

$$\lim_{c\to\infty}\sup_{n\geq 0}\mathbb{E}\Big[\mathbb{1}_{\{|X_n|>c\}}|X_n|\Big] = 0.$$

1) Let us show that if $(X_n)_{n\geq 0}$ is uniformly integrable, then there exists $M > 0$ such that for any $n \geq 0$, we have $\mathbb{E}[|X_n|] \leq M$.

Let us write

$$|X_n| = \mathbb{1}_{\{|X_n|>c\}}|X_n| + \mathbb{1}_{\{|X_n|\leq c\}}|X_n| = A_n + B_n,$$

hence

$$\mathbb{E}[|X_n|] = \mathbb{E}[A_n] + \mathbb{E}[B_n].$$

Because of the uniform integrability, we have

$$\forall \epsilon > 0, \exists\, c(\epsilon) > 0; \forall n \in \mathbb{N}, \mathbb{E}[A_n] \leq \epsilon.$$

For this value of $c(\epsilon)$, for any $n \in \mathbb{N}$, $\mathbb{E}[B_n] \leq c(\epsilon)$. Using $M = \epsilon + c(\epsilon)$, we obtain the required result.

Let $(X_n)_{n\geq 0}$ be an $(\mathcal{F}_n)_{n\geq 0}$-martingale and let T be an $(\mathcal{F}_n)_{n\geq 0}$-stopping time such that $\mathbb{P}(T < \infty) = 1$.

2) Let us show that if the sequence $(X_n)_{n\geq 0}$ is uniformly integrable, then we have $\mathbb{E}[|X_T|] < \infty$ and $\mathbb{E}[X_T] = \mathbb{E}[X_0]$.

According to the previous question, for any $n \geq 0$, we have $\mathbb{E}\Big[|X_n|\Big] \leq M$. On the other hand, since T is almost surely finite, we have $X_{T\wedge n} \to X_T,\ a.s.$ Hence, $|X_{T\wedge n}| \to |X_T|,\ a.s.$ Using Fatou's lemma, it can be concluded that

$$\mathbb{E}[|X_T|] \leq \liminf_{n\to\infty} \mathbb{E}[|X_{T\wedge n}|] \leq M.$$

Since, in addition, $(X_{T\wedge n})$ is a martingale, we have

$$\mathbb{E}[X_{T\wedge n}] = \mathbb{E}[X_{T\wedge 0}] = \mathbb{E}[X_0].$$

From this, it can be derived that $\mathbb{E}[X_T] = \mathbb{E}[X_0]$.

8.4.9. *Exercise 4.9*

Two gamblers Mounir and Benoîte both bet simultaneously on a 1 € coin and the game ends when one of the two is ruined. Mounir initially possesses a € and Benoîte has b €.

1) Let us calculate Mounir's mean fortune at the end of the game.

Mounir's fortune at the instant n is $X_n = a + \sum_{i=1}^{n} Y_i$, where

$$Y_i = \begin{cases} 1\ if & \text{Mounir wins on the } i\text{th turn,} \\ -1\ if & \text{Benoîte wins on the } i\text{th turn.} \end{cases}$$

The sequence (Y_n) is a sequence of centered and i.i.d. random variables. Therefore, (X_n) is an $(\mathcal{F}_n)$-martingale, where $(\mathcal{F}_n)_n$ is the natural filtration of (Y_n).

The time of ruin is $T = \inf\{n \geq 0; X_n = 0 \text{ ou } X_n = a + b\}$. This is a hitting time for the Borel set $\{0, a+b\}$; thus, it is an $(\mathcal{F}_n)$-stopping time.

We have $\mathbb{P}(T < \infty) = 1$. Indeed, we have

$$T > n \Longleftrightarrow 0 < a + \sum_{i=1}^{n} Y_i < a + b$$

$$\Longleftrightarrow -a < \sum_{i=1}^{n} Y_i < b$$

$$\Longleftrightarrow \frac{-a}{\sqrt{n}} < \frac{\sum_{i=1}^{n} Y_i}{\sqrt{n}} < \frac{b}{\sqrt{n}}.$$

Since the sequence (Y_i) is centered and i.i.d., and since $\mathbb{V}(Y_i) = 1$, according to the central limit theorem, the distribution of $\frac{\sum_{i=1}^{n} y_i}{\sqrt{n}}$ converges to the centered, reduced normal distribution when n tends to $+\infty$. In particular, we have

$$\mathbb{P}(T > n) \sim \frac{1}{\sqrt{2\pi}} \int_{-a/\sqrt{n}}^{b/\sqrt{n}} e^{-z^2/2} dz,$$

which tends to 0 when n tends to $+\infty$. From this, it can be derived that

$$\mathbb{P}(T = +\infty) = \mathbb{P}(\cap_{n=1}^{\infty}\{T > n\}) = \lim_{n \to \infty} \mathbb{P}(T > n) = 0.$$

Consequently, $\mathbb{P}(T < \infty) = 1$.

Since we also have $|X_n| \leq a + b$ for any n, according to the stopping theorem,

$$\mathbb{E}[X_T] = \mathbb{E}[X_0] = a.$$

2) Let us calculate the probability that Mounir wins Benoîte's fortune.

We have

$$a = \mathbb{E}[X_T] = 0 \times \mathbb{P}(\text{Mounir ruined}) + (a + b) \times \mathbb{P}(\text{Mounir wins}).$$

Or

$$\mathbb{P}(\text{Mounir ruined}) = 1 - \mathbb{P}(\text{Mounir wins}),$$

we thus obtain

$$\mathbb{P}(\text{Mounir wins}) = \frac{a}{a+b}.$$

8.4.10. *Exercise 4.10*

Let $(\mathcal{F}_n)_{n\geq 0}$ be a filtration and let $(X_n)_{n\geq 0}$ be an $(\mathcal{F}_n)_{n\geq 0}$-martingale.

1) Let us show that if T is a finite $(\mathcal{F}_n)_{n\geq 0}$-stopping time, then X_T is $\mathcal{F}_T$-measurable.

Let B be a Borel set of $\mathbb{R}$. Let us show that $(X_T \in B) \in \mathcal{F}_T$. In order to do this, we must show that $(X_T \in B) \cap (T \leq n) \in \mathcal{F}_n$, for any integer n. This follows immediately from the decomposition

$$(X_T \in B) \cap (T \leq n) = \cup_{i=0}^{n}((X_i \in B) \cap (T = i)) \in \mathcal{F}_n.$$

Let S and T be two $(\mathcal{F}_n)_{n\geq 0}$-stopping times bounded by a constant $c \in \mathbb{N}$ and such that $S \leq T$.

2) Let us show that $\mathcal{F}_S \subset \mathcal{F}_T$.

Let $A \in \mathcal{F}_S$. Therefore, for any n, we have

$$(S \leq n) \cap A \in \mathcal{F}_n.$$

Since $S \leq T$, we have

$$(T \leq n) \cap (S \leq n) = (T \leq n),$$

and consequently, we obtain

$$(T \leq n) \cap A = (S \leq n) \cap A \cap (T \leq n).$$

Since, in addition, $(T \leq n) \in \mathcal{F}_n$, and because T is an $(\mathcal{F}_n)$-stopping time, it can be derived that for any integer n,

$$(T \leq n) \cap A \in \mathcal{F}_n,$$

and consequently, $A \in \mathcal{F}_T$. It thus follows from this that $\mathcal{F}_S \subset \mathcal{F}_T$.

Let $A \in \mathcal{F}_S$ and $R = S\mathbb{1}_A + T\mathbb{1}_{A^c}$.

3) Let us show that R is an $(\mathcal{F}_n)_{n\geq 0}$-stopping time.

We have

$$(R \leq n) = (A \cap (S \leq n)) \cup (A^c \cap (T \leq n)).$$

On the one hand, $(A \cap (S \leq n)) \in \mathcal{F}_n$ according to the definition of $\mathcal{F}_S$. On the other hand, since $A^c \in \mathcal{F}_S \subset \mathcal{F}_T$, we have

$$(A^c \cap (T \leq n)) \in \mathcal{F}_n.$$

Therefore, we do indeed have $(R \leq n) \in \mathcal{F}_n$ and R is an $(\mathcal{F}_n)_{n\geq 0}$-stopping time.

4) Let us show that $\mathbb{E}[X_R] = \mathbb{E}[X_0]$.

Since R is a bounded stopping time, according to the Doob stopping theorem (Theorem 4.3), we directly obtain that $\mathbb{E}[X_R] = \mathbb{E}[X_0]$.

5) Let us show that $\mathbb{E}[X_T \mathbb{1}_A] = \mathbb{E}[X_S \mathbb{1}_A]$.

We have

$$\mathbb{E}[X_T] = \mathbb{E}[X_T \mathbb{1}_A] + \mathbb{E}[X_T \mathbb{1}_A^c] \quad \text{and} \quad \mathbb{E}[X_R] = \mathbb{E}[X_S \mathbb{1}_A] + \mathbb{E}[X_T \mathbb{1}_A^c].$$

By subtracting the two equalities and using the fact that $\mathbb{E}[X_T] = \mathbb{E}[X_R] = \mathbb{E}[X_0]$, we obtain the result

$$\mathbb{E}[X_T \mathbb{1}_A] = \mathbb{E}[X_S \mathbb{1}_A].$$

6) Let us show $\mathbb{E}[X_T | \mathcal{F}_S] = X_S$.

According to Proposition 4.7, the terminal random variable X_S is $\mathcal{F}_S$-measurable. Since, in addition, for any $A \in \mathcal{F}_S$, we have $\mathbb{E}[X_T \mathbb{1}_A] = \mathbb{E}[X_S \mathbb{1}_A]$, it can be derived from this that $\mathbb{E}(X_T | \mathcal{F}_S) = X_S$, from the definition of conditional expectation.

8.4.11. *Exercise 4.11*

Let $(\Omega, \mathcal{F}, \mathbb{P}, (\mathcal{F}_n)_{n\in\mathbb{N}})$ be a filtered probability space, and let $(X_n)_{n\geq 0}$ be $(\mathcal{F}_n)_{n\in\mathbb{N}}$, a positive submartingale. Let $(K_n)_{n\in\mathbb{N}}$ be a positive, bounded and $(\mathcal{F}_n)_{n\in\mathbb{N}}$-predictable process. We consider $((K \cdot X)_n)_{n\in\mathbb{N}}$, the process defined by $(K \cdot X)_0 = 0$ and for any $n \geq 1$,

$$(K \cdot X)_n = \sum_{k=1}^{n} K_k (X_k - X_{k-1}). \qquad (\star)$$

1) Let us show that $((K \cdot X)_n)_{n\in\mathbb{N}}$ is an $(\mathcal{F}_n)_{n\in\mathbb{N}}$-submartingale.

The process $(K \cdot X_n)_{n\in\mathbb{N}}$ is $(\mathcal{F}_n)_{n\in\mathbb{N}}$- adapted since, for any $n \in \mathbb{N}^*$, we have

$$K \cdot X_n = \sum_{k=1}^{n} K_k (X_k - X_{k-1}),$$

is clearly $\mathcal{F}_n$-measurable.

On the other hand, since (K_n) is bounded, there exists $M > 0$ such that

$$\mathbb{E}\left[|K \cdot X_n|\right] \leq M \sum_{k=1}^{n} \mathbb{E}[|X_k - X_{k-1}|] < \infty$$

because (X_n) is integrable.

Finally, for any $n \geq 1$ upon simplification, we obtain

$$\begin{aligned}\mathbb{E}\left[K \cdot X_n - K \cdot X_{n-1}|\mathcal{F}_{n-1}\right] &= \mathbb{E}[K_n(X_n - X_{n-1})|\mathcal{F}_{n-1}] \\ &= K_n\mathbb{E}[X_n - X_{n-1}|\mathcal{F}_{n-1}],\end{aligned}$$

where the final equality is due to the fact that (K_n) is $(\mathcal{F}_n)_{n\in\mathbb{N}}$-predictable. Now, since (X_n) is an $(\mathcal{F}_n)_{n\in\mathbb{N}}$-submartingale, we have

$$\mathbb{E}[X_n - X_{n-1}|\mathcal{F}_{n-1}] \geq 0$$

and consequently

$$\mathbb{E}\left[K \cdot X_n - K \cdot X_{n-1}|\mathcal{F}_{n-1}\right] \geq 0.$$

Let T be an $(\mathcal{F}_n)_{n\in\mathbb{N}}$-stopping time such that there exists $n_0 \in \mathbb{N}^*$; $\forall w \in \Omega, T(w) \leq n_0$.

2) Let us show that $\mathbb{E}[X_0] \leq \mathbb{E}[X_T]$.

Since $(X_{T\wedge n})_{n\in\mathbb{N}}$ is a submartingale, we directly have

$$\mathbb{E}[X_0] = \mathbb{E}[X_{T\wedge 0}] \leq \mathbb{E}[X_{T\wedge n_0}] = \mathbb{E}[X_T].$$

Let $((K \cdot X)_n)_{n\in\mathbb{N}}$ be the sequence of random variables defined by the formula $(\star)$ with $K_k = \mathbb{1}_{\{T\leq k-1\}}$ for any $k \geq 1$.

3) Let us show that $((K \cdot X)_n)_{n\in\mathbb{N}}$ is an $(\mathcal{F}_n)_{n\in\mathbb{N}}$-submartingale.

It is possible to write

$$(K \cdot X)_n = \sum_{l=1}^{n} K_l(X_l - X_{l-1}),$$

where

$$K_l = \mathbb{1}_{T\leq l-1} = \mathbb{1}_{T<l} = 1 - \mathbb{1}_{T\geq l}.$$

The sequence (K_l) is clearly predictable and bounded; therefore, according to Question 1, the process $(K \cdot X)_n$ is a submartingale.

4) Let us show that $\mathbb{E}[X_T] \leq \mathbb{E}(X_{n_0}]$.

Let us take $((K \cdot X)_n)_{n\in\mathbb{N}}$, the sequence defined in the previous question. This is an $(\mathcal{F}_n)_{n\in\mathbb{N}}$-submartingale, and we have

$$(K \cdot X)_{n_0} = \sum_{k=T+1}^{n_0} (X_k - X_{k-1}) = X_{n_0} - X_{T\wedge n_0}.$$

Because $T \wedge n_0 = T$, it follows that

$$\mathbb{E}[X_{n_0}] - \mathbb{E}[X_T] = \mathbb{E}[(K \cdot X)_{n_0}] \geq \mathbb{E}[(K \cdot X)_0] = 0.$$

Let $\lambda > 0$, $A_n := \{\max_{0\leq k\leq n} X_k \geq \lambda\}$ and let T be the function defined on Ω by:

$$T : w \longmapsto T(w) = \inf\{k; , X_k \geq \lambda\} \wedge n.$$

5) Let us show that T is an $(\mathcal{F}_n)_{n\in\mathbb{N}}$-stopping time.

If $k > n$, then we have $(T = k) = \emptyset \in \mathcal{F}_k$.

If $k = n$, then we have $(T = n) = \cap_{i=0}^{n-1}\{X_i' < \lambda\} = A_{n-1}^c \in \mathcal{F}_n$.

And if $k < n$, then we have $(T = k\} = \{X_k \geq \lambda\} \cap (\cap_{i<k}\{X_i < \lambda\}) \in \mathcal{F}_k$.

6) Let us show that $\lambda\mathbb{P}(A_n) \leq \mathbb{E}[X_T \mathbb{1}_{A_n}]$.

We have

$$\lambda\mathbb{P}(A_n) = \mathbb{E}(\lambda\mathbb{1}_{A_n}) \leq \mathbb{E}[X_T \mathbb{1}_{A_n}],$$

where the inequality arises from the fact that $X_T \geq \lambda$ on A_n.

7) Let us show that

$$\mathbb{P}\Big(\max_{0\leq k\leq n} X_k \geq \lambda\Big) \leq \frac{\mathbb{E}[X_n]}{\lambda}.$$

According to Question 6, we have $\lambda\mathbb{P}(A_n) \leq \mathbb{E}[X_T \mathbb{1}_{A_n}]$. Further, $X_T = X_n$ on A_n^c and T is smaller than n; thus, according to Question 4, we also have $\mathbb{E}[X_T) \leq \mathbb{E}[X_n]$. It follows that

$$\mathbb{E}[X_T\mathbb{1}_{A_n}] = \mathbb{E}[X_T] - \mathbb{E}[X_T\mathbb{1}_{A_n^c}] \leq \mathbb{E}[X_n] - \mathbb{E}[X_n\mathbb{1}_{A_n^c}] = \mathbb{E}[X_n\mathbb{1}_{A_n}].$$

Thus, we indeed have $\lambda\mathbb{P}(A_n) \leq \mathbb{E}[X_n\mathbb{1}_{A_n}] \leq \mathbb{E}[X_n]$.

8.5. Solutions to exercises in Chapter 5

8.5.1. *Exercise 5.1*

Consider the following market model on $\Omega = (\omega_1, \omega_2)$:

	$n = 0 : (S_0^0, S_0^1)$	$n = 1 : (S_1^0, S_1^0)$
ω_1	$(1, 2)$	$(1.05, 2.5)$
ω_2	$(1, 2)$	$(1.05, 5)$

1) The event tree representing the evolution of this financial market is given in Figure 8.3.

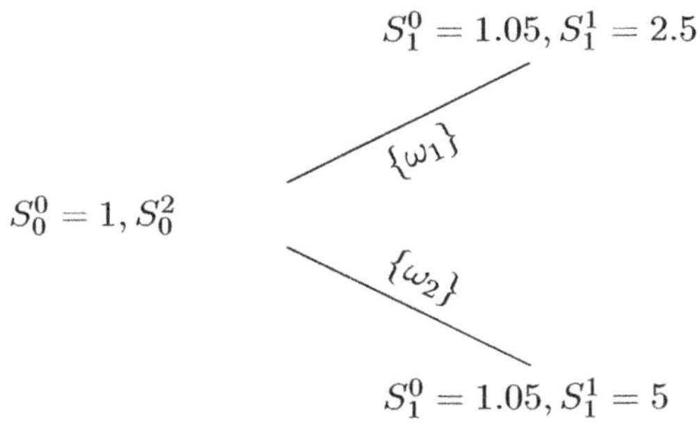

Figure 8.3. *Event tree for the financial market in Exercise 5.1*

2) On reading the tree, it can be seen that the σ-algebras that represent the information available at each instant are $\mathcal{F}_0 = \{\emptyset, \Omega\}$ and $\mathcal{F}_1 = \mathcal{P}(\Omega)$.

3) The interest rate for the risk-free asset is the solution to the equation

$$S_0^0(1 + r) = S_1^0.$$

We thus obtain $r = 5\%$.

4) Let us calculate the acquisition cost of the initial portfolio of an investor short selling four shares in a risk-free asset (i.e. borrowing the risk-free asset) to buy two shares in a risky asset. We thus have

$$V_0 = -4S_0^0 + 2S_0^1 = -4 + 4 = 0.$$

The acquisition cost of the portfolio is zero.

5) Let us calculate the possible values of the portfolio after the evolution of the asset values and after reimbursing the borrowed risk-free assets.
For the hazard ω_1, we have

$$V_1(\omega_1) = -4S_1^0(\omega_1) + 2S_1^1(\omega_1) = -4 \times 1.05 + 2 \times 2.5 = -4.2 + 5 = 0.8,$$

for the hazard ω_2, we have

$$V_1(\omega_2) = -4S_1^0(\omega_2) + 2S_1^1(\omega_2) = -4 \times 1.05 + 2 \times 5 = -4.2 + 10 = 5.8.$$

In all the cases, the final wealth is strictly positive.

6) Since we have just constructed an arbitrage opportunity, this market is not viable.

8.5.2. *Exercise 5.2*

Consider the following market model on $\Omega = \{\omega_1, \omega_2\}$

	$n = 0 : (S_0^0, S_0^1)$	$n = 1 : (S_1^0, S_1^1)$	$n = 2 : (S_2^0, S_2^1)$
ω_1	$(1, 1)$	$(1.1, 1.2)$	$(1.2, 1.5)$
ω_2	$(1, 1)$	$(1.1, 1.5)$	$(1.2, 1.2)$

1) Consider the strategy $\Phi_1(\omega_1) = \Phi_1(\omega_2) = (-1, 2)$, $\Phi_2(\omega_1) = \Phi_2(\omega_2) = (-1, 1)$. Let us calculate the wealth at every instant corresponding to this strategy. We have

$$\begin{aligned}
V_0 &= -1S_0^0 + 2S_0^1 = 1 \\
V_1(\omega_1) &= -1.1 + 2 \times 1.2 = 1.3 \quad \text{before redistribution of the portfolio} \\
V_1(\omega_2) &= -1.1 + 2 \times 1.5 = 1.9 \quad \text{before redistribution of the portfolio} \\
V_1(\omega_1) &= -1.1 + 1.2 = 0.1 \quad \text{after redistribution of the portfolio} \\
V_1(\omega_2) &= -1.1 + 1.5 = 0.4 \quad \text{after redistribution of the portfolio} \\
V_2(\omega_1) &= -1.2 + 1.5 = 0.3 \\
V_2(\omega_2) &= -1.2 + 1.2 = 0.
\end{aligned}$$

This strategy is indeed predictable, it gives a non-negative wealth at each instant. On the other hand, it is not self-financed as the value of V_1 decreases upon redistribution of the portfolio: money is withdrawn. This strategy is thus not admissible.

2) Consider the strategy $\Phi_1(\omega_1) = \Phi_1(\omega_2) = (-1, 2)$, $\Phi_2(\omega_1) = \Phi_2(\omega_2) = (-2, 4)$. Let us calculate the wealth at every instant corresponding to this strategy. We have

$$V_0 = -1S_0^0 + 2S_0^1 = 1$$

$$V_1(\omega_1) = -1.1 + 2 \times 1.2 = 1.3 \quad \text{before redistribution of the portfolio}$$

$$V_1(\omega_2) = -1.1 + 2 \times 1.5 = 1.9 \quad \text{before redistribution of the portfolio}$$

$$V_1(\omega_1) = -2 \times 1.1 + 4 \times 1.2 = 2.6 \quad \text{after redistribution of the portfolio}$$

$$V_1(\omega_2) = -2 \times 1.1 + 4 \times 1.5 = 3.8 \quad \text{after redistribution of the portfolio}$$

$$V_2(\omega_1) = -2 \times 1.2 + 4 \times 1.5 = 3.6$$

$$V_2(\omega_2) = -2 \times 1.2 + 4 \times 1.2 = 2.4.$$

This strategy is indeed predictable; it gives a non-negative wealth at each instant. On the other hand, it is not self-financed as the value of V_1 increases upon the redistribution of the portfolio: money is added. This strategy is thus not admissible.

3) Consider the strategy $\Phi_1(\omega_1) = \Phi_1(\omega_2) = (-1, 1)$, $\Phi_2(\omega_1) = (0.08, 0.01)$, $\Phi_2(\omega_2) = (-16, 12)$. Let us calculate the wealth at every instant corresponding to this strategy. We have

$$V_0 = S_0^0 + S_0^1 = 0$$

$$V_1(\omega_1) = -1.1 + 1.2 = 0.1 \quad \text{before redistribution of the portfolio}$$

$$V_1(\omega_2) = -1.1 + 1.5 = 0.4 \quad \text{after redistribution of the portfolio}$$

$$V_1(\omega_1) = 0.08 \times 1.1 + 0.01 \times 1.2 = 0.1 \quad \text{after redistribution of the portfolio}$$

$$V_1(\omega_2) = -16 \times 1.1 + 12 \times 1.5 = 0.4 \quad \text{after redistribution of the portfolio}$$

$$V_2(\omega_1) = 0.08 \times 1.2 + 0.01 \times 1.5 = 0.111$$

$$V_2(\omega_2) = -16 \times 1.2 + 12 \times 1.2 = -4.8.$$

This strategy is indeed predictable and self-financed. However, it gives the portfolio a negative value at time 2 for the hazard ω_2. This strategy is thus not admissible.

4) Consider the strategy $\Phi_1(\omega_1) = \Phi_1(\omega_2) = (-2, 2)$, $\Phi_2(\omega_1) = \Phi_2(\omega_2) = (-2, 2)$. Let us calculate the wealth at every instant corresponding to this strategy. We have

$$V_0 = -2S_0^0 + 2S_0^1 = 0$$

$$V_1(\omega_1) = -2 \times 1.1 + 2 \times 1.2 = 0.2 \quad \text{before and after redistribution of the portfolio}$$

$$V_1(\omega_2) = -2 \times 1.1 + 2 \times 1.5 = 0.8 \quad \text{before redistribution of the portfolio}$$

$$V_2(\omega_1) = -2 \times 1.2 + 2 \times 1.5 = 0.6$$

$$V_2(\omega_2) = -2 \times 1.2 + 2 \times 1.2 = 0.$$

This strategy is indeed predictable, self-financed and gives a non-negative wealth at each instant. This is thus an admissible strategy.

6) The final strategy is an arbitrage opportunity. This market is thus not viable.

8.5.3. *Exercise 5.3*

Consider the following market model on $\Omega = (\omega_1, \omega_2, \omega_3)$:

	$n = 0 : (S_0^0, S_0^1, S_0^2)$	$n = 1 : (S_1^0, S_1^1, S_1^2)$
ω_1	$(1, 1, 1)$	$(1.1, 1.65, 0.88)$
ω_2	$(1, 1, 1)$	$(1.1, 1.65, 0.88)$
ω_3	$(1, 1, 1)$	$(1.1, 0.55, 1.32)$

1) The event tree representing the evolution of this financial market is given in Figure 8.4.

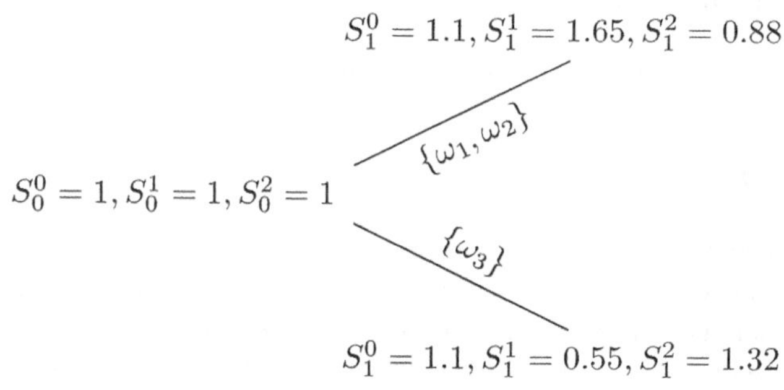

Figure 8.4. *Event tree for the financial market in Exercise 5.3*

2) On reading the tree, it is seen that the σ-algebras representing the information available at each instant are $\mathcal{F}_0 = \{\emptyset, \Omega\}$ and $\mathcal{F}_1 = \{\emptyset, \{\omega_1, \omega_2\}, \{\omega_3\}, \Omega\}$.

3) Let us show that the probability $\mathbb{P}_1(\omega_1) = 1/4$, $\mathbb{P}_1(\omega_2) = 1/4$, $\mathbb{P}_1(\omega_3) = 1/2$ is a risk-neutral probability.

The values of the discounted risky assets are

	$n = 0\ :\ (\widetilde{S}_0^1, \widetilde{S}_0^2)$	$n = 1\ :\ (\widetilde{S}_1^1, \widetilde{S}_1^2)$
ω_1	$(1, 1)$	$(1.5, 0.8)$
ω_2	$(1, 1)$	$(1.5, 0.8)$
ω_3	$(1, 1)$	$(0.5, 1.2)$

These are well adapted with respect to the filtration $(\mathcal{F}_0, \mathcal{F}_1)$ and integrable. Let us calculate the conditional expectations under $\mathbb{P}_1$:

$$\mathbb{E}_1[\widetilde{S}_1^1 | \mathcal{F}_0] = \mathbb{E}_1[\widetilde{S}_1^1] = 1.5/4 + 1.5/4 + 0.5/2 = 1 = \widetilde{S}_0^1,$$
$$\mathbb{E}_1[\widetilde{S}_1^2 | \mathcal{F}_0] = \mathbb{E}_1[\widetilde{S}_1^1] = 0.8/4 + 0.8/4 + 1.2/2 = 1 = \widetilde{S}_0^2.$$

The calculation here is a simple one of expectation as only the constants are $\mathcal{F}_0$-measurable. We thus do have martingales, $\mathbb{P}_1$ is a risk-neutral probability.

4) Let us show that the probability $\mathbb{P}_2(\omega_1) = 2/10, \mathbb{P}_2(\omega_2) = 3/10, \mathbb{P}_2(\omega_3) = 1/2$ is a risk-neutral probability.

The discounted risky assets are unchanged and, therefore, adapted to the filtration $(\mathcal{F}_0, \mathcal{F}_1)$ and integrable. Let us calculate the conditional expectations under $\mathbb{P}_2$:

$$\mathbb{E}_2[\widetilde{S}_1^1 | \mathcal{F}_0] = \mathbb{E}_2[\widetilde{S}_1^1] = 1.5 \times 0.2 + 1.5 \times 0.3 + 0.5/2 = 1 = \widetilde{S}_0^1,$$
$$\mathbb{E}_2[\widetilde{S}_1^2 | \mathcal{F}_0] = \mathbb{E}_2[\widetilde{S}_1^1] = 0.8 \times 0.2 + 0.8 \times 0.3 + 1.2/2 = 1 = \widetilde{S}_0^2,$$

we thus again have martingales.

5) The market is viable because we found risk-neutral probabilities.

8.5.4. *Exercise 5.4*

We wish to construct an arbitrage opportunity in a market following the Cox, Ross and Rubinstein model with $d \geq r$. Since we have $u > d \geq r$, this signifies that it is always more advantageous to invest in a risky asset than the risk-free asset. This leads us to the following proposition: borrowing a share in a risk-free asset to invest it in the risky asset: $\Phi_1 = (-1, 1)$. We thus have $V_0 = -1 + 1 = 0$ and $V_1 = -(1+r) + 1 + d \geq 0$ on one ω_i (hazard) and $V_1 = -(1+r) + 1 + u > 0$ on another, because $u > d$. This is thus an arbitrage opportunity.

8.5.5. *Exercise 5.5*

We wish to construct an arbitrage opportunity in a market following the Cox, Ross and Rubinstein model with $u \leq r$. Since we have $r \geq u > d$, this means that it is more advantageous to invest in the risk-free asset than in the risky asset. This leads us to the following proposition: borrowing a share in a risky asset to invest in the risk-free asset: $\Phi_1 = (1, -1)$. We then have $V_0 = 1 - 1 = 0$ and $V_1 = 1 + r - (1 + d) = r - d > r - u \geq 0$ on one ω_i (hazard) and $V_1 = 1 + r - (1 + u) \geq 0$ on the other. This is thus an arbitrage opportunity.

8.6. Solutions to the practical exercises in Chapter 5

The solutions are given in the language R.

8.6.1. *Practical exercise 5.6.1*

1) Let us construct a trajectory simulator for the risky asset in the Cox, Ross and Rubinstein model with the additional assumption that the sequence (T_n) is iid. We write

$$\mathbb{P}(T_1 = 1 + u) = 1 - \mathbb{P}(T_1 = 1 + d) = q.$$

The simulator takes as inputs the parameters u, d and q for the model and the simulation horizon N. It returns the vector containing the values $(S_n^1)_{0 \leq n \leq N}$. In order to do this, we simulate the i.i.d. sequence $(T_n)_{1 \leq n \leq N}$ and calculate the cumulative sums. This calculation may be carried out directly without any loop.

```
CRR <- function(u,d,q,N){
  T<- sample(c(1+u,1+d), size=N, replace=TRUE, prob=c(q, 1-q))
  S<-c(1,cumprod(T))
  return(S)
}
```

2) Let us graphically represent some trajectories under the risk-neutral probability for $r = 0$, $u = 10\%$, $d = -5\%$, $q = 1/3$ and $N = 50$. In order to do this, we call up the function created earlier. The outputs are given in Figure 8.5.

```
u <- 0.1
d <- -0.05
N <- 50
q <- 1/3
traj <- CRR(u,d,q,N)
plot(0:N, traj ,type="b")
```

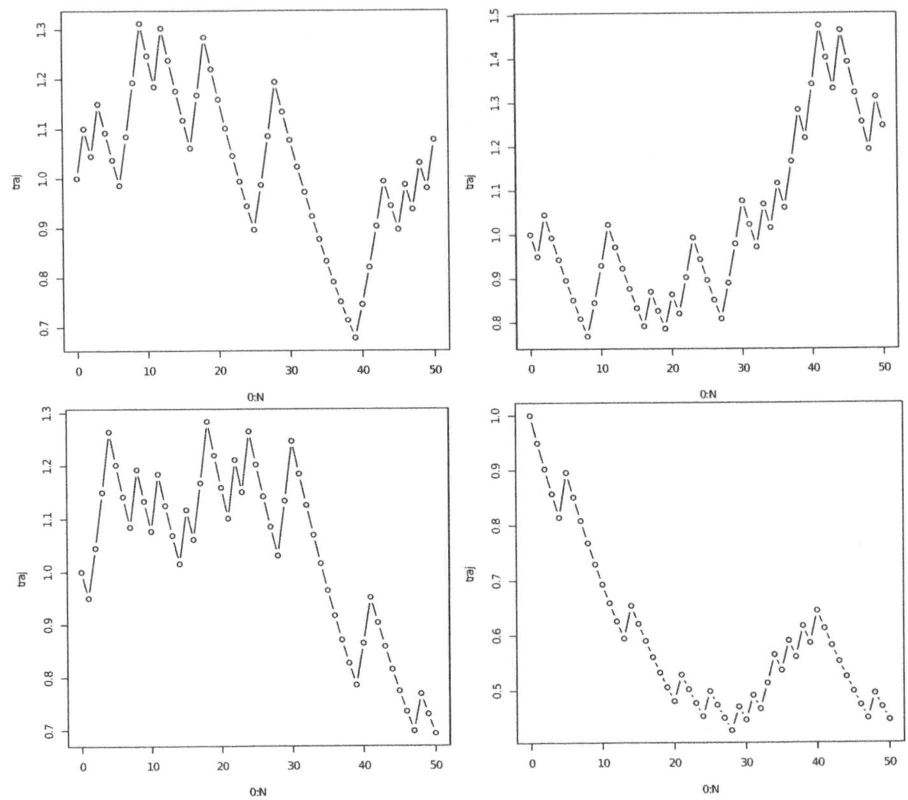

Figure 8.5. *Trajectories of the risky asset for the Cox, Ross and Rubinstein model with parameters* $r = 0$, $u = 10\%$, $d = -5\%$, $q = 1/3$ *for a horizon horizon* $N = 50$

8.6.2. *Practical exercise 5.6.2*

We examine a portfolio optimization problem in the Cox, Ross and Rubinstein model.

Let V_0 be an investor's wealth at time 0. The investor can invest their money either in a risky asset or in a risk-free asset, following an admissible strategy. We use ϕ_n^0 and ϕ_n^1 to denote the number of shares in the risk-free asset and in the risky asset held between the instants $n-1$ and n. For problems of this type, it is easier to manipulate the proportions of wealth invested in the risky asset rather than directly manipulating the number of shares. Let π_n be the *proportion* of the wealth invested in the risky asset between the instants $n-1$ and n, i.e.,

$$\pi_n = \frac{\phi_n^1 S_{n-1}^1}{V_{n-1}}.$$

1) Let us express ϕ_n^0 and ϕ_n^1 as functions of π_n, S_{n-1}^0, S_{n-1}^1 and V_{n-1} for any n.

According to the formula, we have

$$\phi_n^1 = \frac{\pi_n V_{n-1}}{S_{n-1}^1},$$

further, the wealth at the instant $n-1$ can be broken up as

$$V_{n-1} = \phi_n^0 S_{n-1}^0 + \phi_n^1 S_{n-1}^1.$$

Therefore, we have

$$\phi_n^0 = \frac{V_{n-1} - \phi_n^1 S_{n-1}^1}{S_{n-1}^0} = \frac{V_{n-1} - \pi_n V_{n-1}}{S_{n-1}^0} = \frac{(1-\pi_n)V_{n-1}}{S_{n-1}^0}.$$

2) Thus, we can rewrite the expression for the wealth at the time n using only the proportion π_n. We obtain

$$\begin{aligned} V_n &= \phi_n^0 S_n^0 + \phi_n^1 S_n^1 \\ &= \frac{(1-\pi_n)V_{n-1}}{S_{n-1}^0} S_n^0 + \frac{\pi_n V_{n-1}}{S_{n-1}^1} S_n^1 \\ &= V_{n-1}\Big((1-\pi_n)\frac{S_n^0}{S_{n-1}^0} + \pi_n \frac{S_n^1}{S_{n-1}^1}\Big) = V_{n-1}\big((1-\pi_n)(1+r) + \pi_n T_n\big). \end{aligned}$$

This formulation in terms of proportion makes it possible to use the multiplicative nature of the asset dynamic. We thus also arrive once more at a multiplicative formulation for the wealth.

3) We now wish to represent on the same graph and for the **same random sampled trajectory**, the evolution of the value of the risky asset, the evolution of the value of the risk-free asset and the evolution of the wealth for the two following strategies:

1) The *proportion* of the wealth invested in the risky asset is fixed over time, equal to $1/4$.

2) The *proportion* of the wealth invested in the risky asset only takes the value 0 or 1. It takes the value 1 when the price of the risky asset strictly exceeds that of the risk-free asset, and the value 0 when the risky asset is lower than or equal to the risk-free asset, *while remaining predictable*.

We take the following parameters: initial wealth $V_0 = 1$ the risky asset evolves as per the Cox, Ross and Rubinstein model with parameters $d = -2\%$, $u = 10\%$, $q = 0.52$, interest rate $r = 4\%$ and duration of investment: $N = 100$ periods.

We must be careful to use the same random sampling trajectories of the variables T_n for all three strategies; otherwise, we cannot make direct comparisons. This is the script.

```
r <- 0.04
u <- 0.1
d <- -0.02
N <- 100
q <- 0.52
## risk-free asset
S0 <- (1+r)^(0:N)
## risky asset
T<- sample(c(1+u,1+d), size=N, replace=TRUE, prob=c(q, 1-q))
S<-c(1,cumprod(T))
## portfolio strategy 1/4
pi <- 1/4
Vquart <- cumprod(c(1, ((1-pi)*(1+r)+pi*T)))
## portfolio threshold
Vs <- NULL
Vs[1] <- 1
for (i in 1:N){
        if (S[i]>S0[i]){
                prop <- 1
        }
        else{
                prop <- 0
        }
        Vs[i+1]<- Vs[i]*((1-prop)*(1+r)+prop*T[i])
}
mat <- cbind(S0,S,Vquart,Vs)
matplot(0:N, mat, type=c("l"), col=c("grey", "blue", "magenta",
    "red"))
```

The graphical outputs are given in Figure 8.6. It can be seen that according to the sampled trajectories, the final wealth is greatest for one or the other of the two strategies. We now consider that the investor's utility function is logarithmic, and we wish to find the strategy $\pi^* = (\pi_n^*)_{1\leq n\leq N}$ that maximizes the expectation of the utility of the wealth at the maturity date N:

$$\sup_{\pi \text{ admissible}} \mathbb{E}[\log V_N(\pi)].$$

As the set of admissible policies is infinite, it is not evident in most cases that such an optimal policy exists. In this specific case, it exists and it may be explicitly calculated by making use of the fact that the logarithmic function behaves well with respect to the product. For example, it is possible to obtain the expression for the optimal policy by carrying out a recurrence on the horizon N. It is accepted that the optimal strategy is constant over time and maximizes the expression

$$q\log(\pi^*(u-r)+1+r)+(1-q)\log(\pi^*(d-r)+1+r). \qquad [8.3]$$

We wish to compare the performances of strategies 1 and 2, given above, and the performance of the optimal strategy. We will use the same parameters as in the previous question.

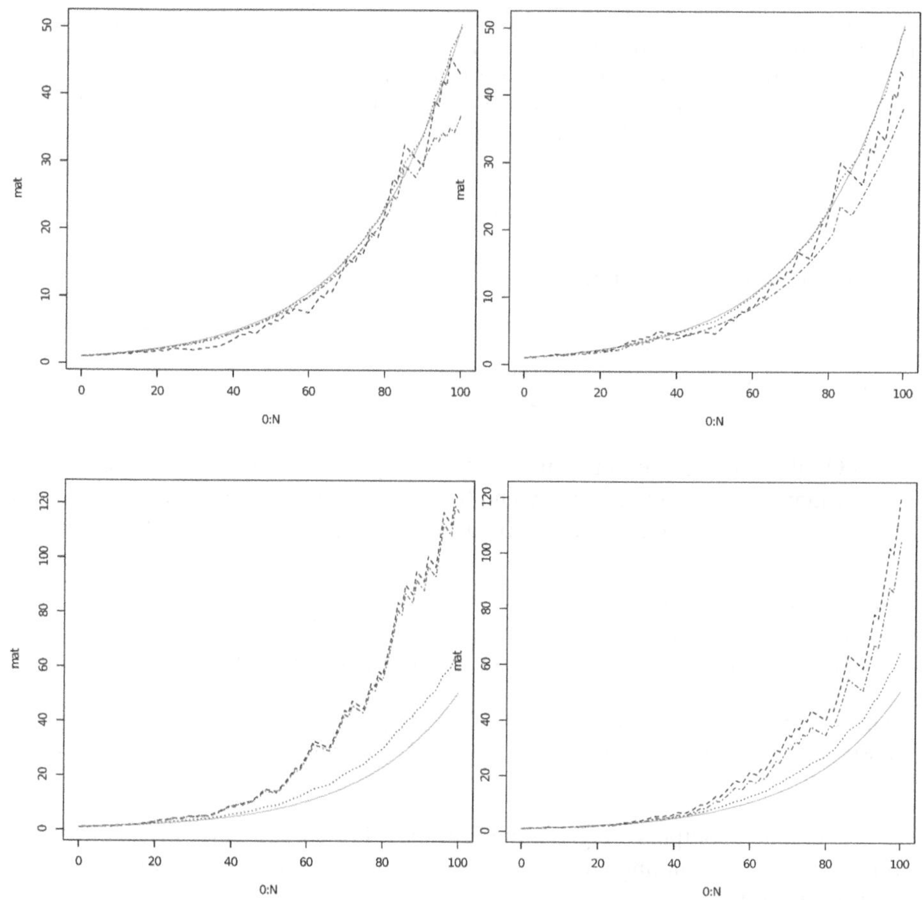

Figure 8.6. *Trajectories of the risky asset (blue) and the risk-free asset (gray) and wealth for the $1/4$ strategy (magenta) and $0-1$ strategy (red) for the Cox, Ross and Rubinstein model with parameters $r = 4\%$, $d = -2\%$, $u = 10\%$, $q = 0.52$ and duration of investment $N = 100$ periods. For a color version of this figure, see www.iste.co.uk/desaporta/martingales.zip*

4) Let us first calculate the optimal proportion π^*. There are several ways of doing this. For example, it is possible to explicitly calculate the derivative of expression [8.3] and then analytically find the point where it is canceled out. It is also possible to numerically find the maximum by calculating a sufficiently fine discretization of expression [8.3], or again, combining the two approaches by analytically calculating the derivative and then use a dichotomy method or gradient method to numerically calculate its 0. Below we have given the code for the discretization approach.

```
optimal <- function(u,d,r,q){
        prop <- seq(0,1, by=0.00001)
        x <- q*log(prop*(u-r)+1+r)+(1-q)*log(prop*(d-r)+1+r)
        ind <- which.max(x)
        return(prop[ind])
}
```

For our parameters, we find

```
> optimal(u,d,r,q)
[1] 0.69333
```

5) On the same graph, let us trace a trajectory for the *logarithm* of the wealth at each instant for the optimal strategy and for strategies 1 and 2. We once again use the code from Question 3 and add the optimal strategy and the shift to the logarithm.

```
## optimal wealth
pstar <- optimal(u,d,r,q)
Vstar <- cumprod(c(1, ((1-pstar)*(1+r)+pstar*T)))

mat <- cbind(log(Vquart),log(Vs),log(Vstar))
matplot(0:N, mat, type=c("l"), col=c("blue", "black", "red"))
```

The graphical outputs are given in Figure 8.7. For certain trajectories, the optimal strategy is not the winning strategy. This is because we have optimized on average, not almost surely.

6) On the same graph, let us trace the expectation of the *logarithm* of the wealth at each instant, for each of the three strategies. We calculate the expectation using the Monte Carlo method.

```
r <- 0.04
u <- 0.1
d <- -0.02
N <- 100
q <- 0.52
pstar <- optimale(u,d,r,q)
S0 <- (1+r)^(0:N)

Vquart <- matrix(0,N+1,10000)
Vs <- matrix(0,N+1,10000)
Vstar <- matrix(0,N+1,10000)

for (mc in 1:1000){
        T<- sample(c(1+u,1+d), size=N, replace=TRUE,
    prob=c(q, 1-q))
        S<-c(1,cumprod(T))
        Vquart[,mc] <- cumprod(c(1, ((1-pi)*(1+r)+pi*T)))
        Vs[1,mc] <- 1
                for (i in 1:N){
                        if (S[i]>S0[i]){
                                prop <- 1
                        }
                        else{
                                prop <- 0
                        }
                Vs[i+1,mc]<- Vs[i,mc]*((1-prop)*(1+r)
     +prop*T[i])
                }
        Vstar[,mc] <- cumprod(c(1, ((1-pstar)*(1+r)+pstar*T)))
}
mVquart = apply(log(Vquart),1,sum)/10000
mVs = apply(log(Vs),1,sum)/10000
mVstar = apply(log(Vstar),1,sum)/10000

mat <- cbind(mVquart,mVs,mVstar)
matplot(0:N, mat, type=c("l"), col=c("blue", "black", "red"))
```

The graphical outputs are given in Figure 8.8. There are very small differences; however, it is still possible to see that the optimal strategy is the best performing strategy. Let us verify this by looking at the approximation of the expectation of the terminal wealth for each strategy. Let us verify this by looking at the approximation for the expectation of terminal wealth for each strategy.

```
VVq=log(Vquart[N+1,])
VVs=log(Vs[N+1,])
VVst=log(Vstar[N+1,])

ICq=c(mean(VVq)-1.96*sqrt(var(VVq)/10000), mean(VVq),
    mean(VVq)+1.96*sqrt(var(VVq)/10000))
ICs=c(mean(VVs)-1.96*sqrt(var(VVs)/10000), mean(VVs),
    mean(VVs)+1.96*sqrt(var(VVs)/10000))
ICst=c(mean(VVst)-1.96*sqrt(var(VVst)/10000), mean(VVst),
    mean(VVst)+1.96*sqrt(var(VVst)/10000))
```

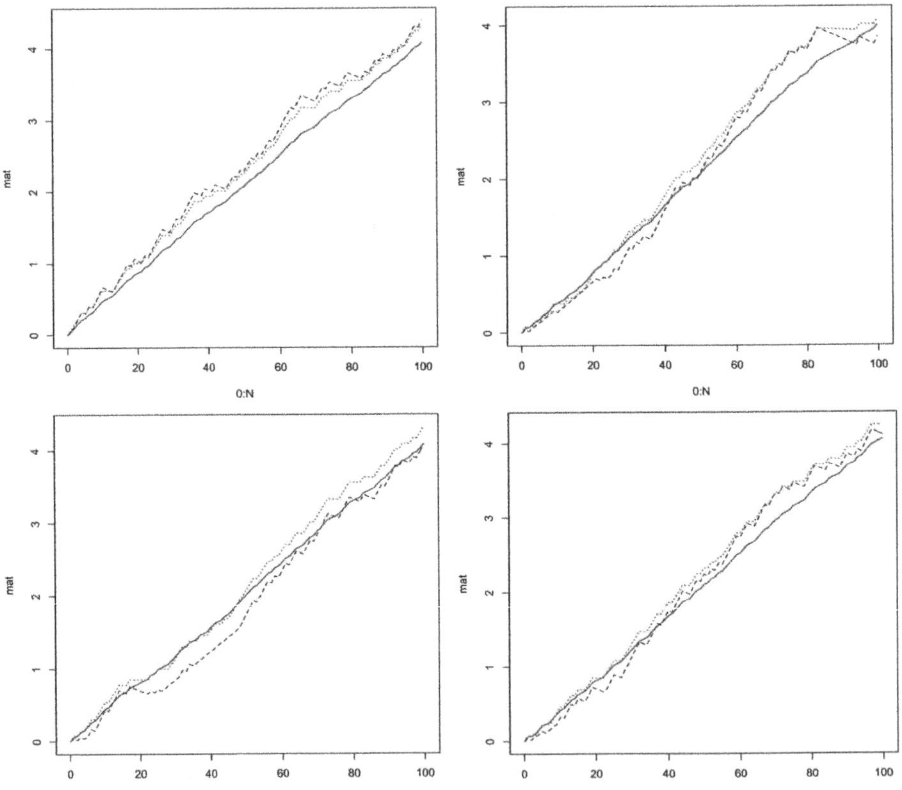

Figure 8.7. *Trajectory of the logarithm of the wealth for the optimal strategy (red) $1/4$ strategy (blue) and $0-1$ strategy (black) for the Cox, Ross and Rubinstein model with parameters $r = 4\%$, $d = -2\%$, $u = 10\%$, $q = 0.52$ and duration of investment $N = 100$ periods. For a color version of this figure, see www.iste.co.uk/desaporta/martingales.zip*

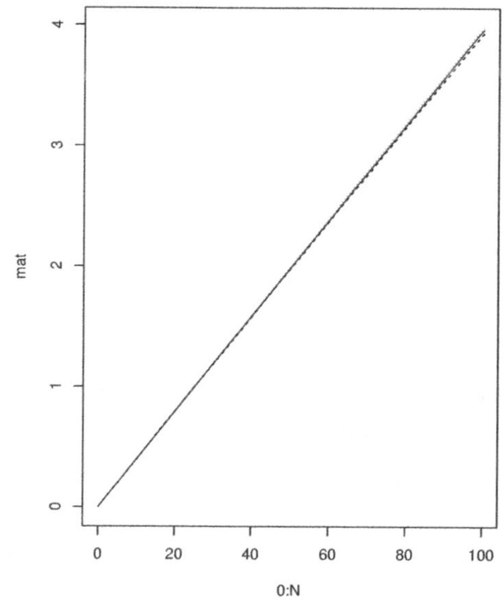

Figure 8.8. *Trajectories of the expectation of the logarithm of the wealth for the optimal strategy (red), $1/4$ strategy (blue) and $0-1$ strategy (black) for the Cox, Ross and Rubinstein model with parameters $r = 4\%$, $d = -2\%$, $u = 10\%$, $q = 0.52$ and duration of investment $N = 100$ periods*

We obtain the following values (with the confidence interval at 95%):

– Optimal strategy: 4.002 ($[3.994; 4.010]$).

– Strategy $1/4$: 3.969 ($[3.966; 3.972]$).

– Strategy $0-1$: 3.955 ($[3.946; 3.963]$).

The confidence intervals do not intersect; it is possible to say with 95% certainty that the optimal strategy is the best and that the $0-1$ strategy is the poorest.

8.6.3. *Practical exercise 5.6.3*

In this section, the investor is allowed to withdraw a *proportion* c_n of their wealth at the instant n, *after* updating the asset prices but *before* the redistribution of the portfolio for the next period. Thus, they only reinvest the non-withdrawn part. The corresponding investment-withdrawal strategy (π_n, c_n) is no longer self-financing, but it must remain predictable and the wealth, after withdrawal, must be positive or zero at each instant in order for the strategy to be considered admissible. Thus, it is possible to show that the new wealth at the time n, after the change in asset prices and after withdrawal has the value:

$$V_n(\pi, c) = \prod_{i=1}^{n}(1 - c_i)\big(\pi_i T_i + (1 - \pi_i)(1 + r)\big),$$

such that the value of the wealth withdrawn at the instant n is $R_n(\pi, c)$ with

$$R_n(\pi, c) = \begin{cases} \frac{c_n V_n(\pi,c)}{1-c_n} & \text{if } c_n \neq 1, \\ V_{n-1}(\pi, c)\big(\pi_n T_n + (1 - \pi_n)(1 + r)\big) & \text{otherwise.} \end{cases}$$

1) We wish to graphically represent the evolution of the wealth for the following investment strategy with withdrawal:

a) The proportion of the wealth invested in the risky asset is fixed over time, equal to $1/4$.

b) The proportion of the wealth withdrawn at each instant is equal to 1% on the interval $[1, 80]$, 5% on the interval $[80, 90]$, 10% on the interval $[90, 95]$, 25% on the interval $[95, 100]$ and on the date of maturity, all the remaining wealth is withdrawn.

We take the following parameters: initial wealth $V_0 = 1$, the risky asset evolves following the Cox, Ross and Rubinstein model with parameters $d = -2\%$, $u = 10\%$, $q = 0.52$, interest rate $r = 4\%$ and duration of investment: $N = 100$ periods.

We adapt the code from the previous exercise to take into account the withdrawal.

```
r <- 0.04
u <- 0.1
d <- -0.02
N <- 100
q <- 0.52
pi <- 1/4
conso <- c(rep(0.01,80),rep(0.05,10),rep(0.1,5),rep(0.25,4),1)
V <- NULL
V[1] <- 1
## risky asset
T<- sample(c(1+u,1+d), size=N, replace=TRUE, prob=c(q, 1-q))
## portfolio
V[2:(N+1)] <- cumprod((1-conso)*((1-pi)*(1+r)+pi*T))
plot(0:N,V,type="b")
```

The graphical outputs are given in Figure 8.9. The wealth increases when the withdrawals are small and then falls drastically as the withdrawal amount increases. We once again now consider that the investor's utility function is logarithmic, and we wish to find the investment strategy with withdrawal (π_n, c_n) that maximizes the expectation of the cumulative sum of the withdrawal utility up to the date of maturity N:

$$\max_{(\pi_n, c_n)} \mathbb{E}\Big[\sum_{n=1}^{N} \log\left(R_n(\pi, c)\right)\Big].$$

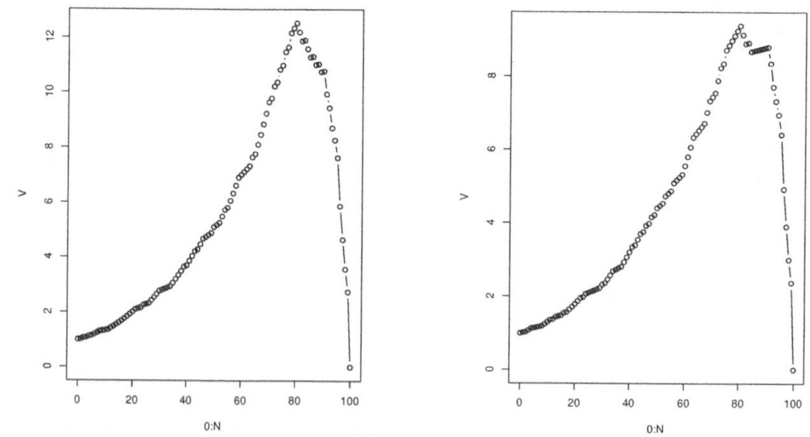

Figure 8.9. *Trajectories of the wealth for the investment-withdrawal strategy in Exercise 5.6.3, Question 1, in the Cox, Ross and Rubinstein model with parameters* $r = 4\%$, $d = -2\%$, $u = 10\%$, $q = 0.52$ *and duration of investment* $N = 100$ *periods*

Once again, in most cases, an optimal strategy does not necessarily exist. In this specific case, it exists and we can still explicitly calculate it by using the multiplicative properties of the logarithmic function. For instance, we can use an induction on N to calculate this strategy. It is accepted that the investment strategy with optimal withdrawal is given by

– $\pi_n = \pi^*$ for any $1 \leq n \leq N$;

– $c_n = \frac{1}{N+1-n}$ for any $1 \leq n \leq N$,

with the same π^* as in the previous exercise.

We wish to compare the performances of the strategy from the above exercise and that of the optimal strategy. We will use the same parameters as for the previous exercise.

2) Let us start by writing a function `strategies_conso` that calculates, for the *same* random sampled trajectory, the wealth at any instant for the optimal strategy and the above strategy.

```
strategie_conso <- function(r,u,d,N,q){
  pi <- 1/4
  conso <- c(rep(0.01,80),rep(0.05,10),rep(0.1,5),
     rep(0.25,4),1)
  pistar <- optimal(u,d,r,q)
  cstar <- 1/(N:1)
  V <- NULL
  V[1] <- 1
  Vstar <- NULL
  Vstar[1] <- 1
## risky asset
  T<- sample(c(1+u,1+d), size=N, replace=TRUE, prob=c(q, 1-q))
## portfolio
  V[2:(N+1)] <- cumprod((1-conso)*((1-pi)*(1+r)+pi*T))
## optimal portfolio
  Vstar[2:(N+1)] <- cumprod((1-cstar)*((1-pistar)*(1+r)
    +pistar*T))
  mat <- cbind(V, Vstar)
  return(mat)
}
```

3) On the same graph, let us trace a trajectory for the wealth at each instant for each of the two strategies. The outputs are given in Figure 8.10. The growth of wealth is smaller for the optimal strategy. Let us recall that the objective is not to maximize the wealth but to maximize the withdrawal utility. Thus, this is not necessarily a cause for concern.

4) For the same sampled trajectories and for each strategy, let us now calculate, at every instant n, the value of the cumulative sum of the logarithm of withdrawal, up to the instant n.

```
pi <- 1/4
conso <- c(rep(0.01,80),rep(0.05,10),rep(0.1,5),rep(0.25,4),1)
pistar <- optimal(u,d,r,q)
cstar <- 1/(N:1)

mat <- strategie_conso(r,u,d,N,q)
matplot(0:N, mat, type=c("l"), col=c("blue", "red"))
R <- c(mat[2:N,1]*conso[1:(N-1)]/(1-conso[1:(N-1)]),mat[N,1])
Rstar<- c(mat[2:N,2]*cstar[1:(N-1)]/(1-cstar[1:(N-1)]),
     mat[N,2])
lR <- cumsum(log(R))
lRstar <- cumsum(log(Rstar))
matR <- cbind(lR,lRstar)
matplot(1:N,matR, type=c("l"), col=c("blue", "red"))
```

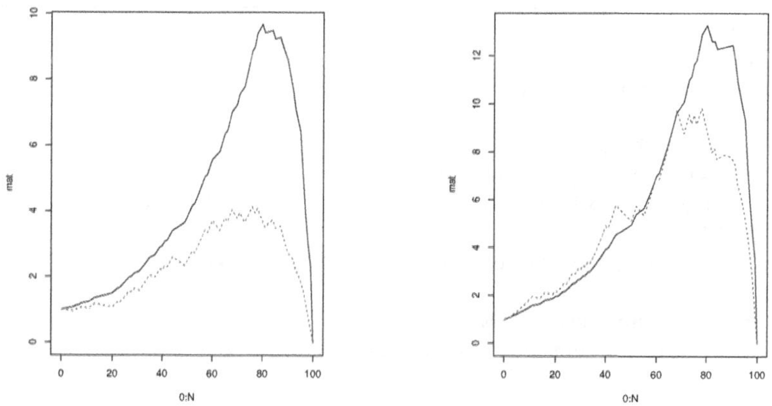

Figure 8.10. *Trajectories of the wealth for the investment-withdrawal strategy in Exercise 5.6.3, Question 1 (blue) and the optimal strategy (red), in the Cox, Ross and Rubinstein model with the parameters* $r = 4\%$, $d = -2\%$, $u = 10\%$, $q = 0.52$ *and duration of investment* $N = 100$ *periods. For a color version of this figure, see www.iste.co.uk/desaporta/martingales.zip*

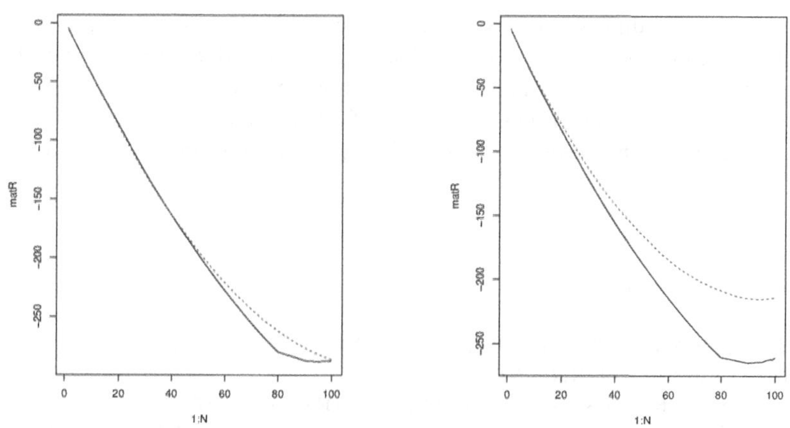

Figure 8.11. *Trajectories of the logarithm of the cumulative withdrawal for the wealth trajectories in Figure 8.10*

We obtain the graphs in Figure 8.11. In the first figure, the curves are quite close; in the second, the optimally controlled wealth gives a withdrawal utility that is clearly higher than that in the other strategy.

5) Let us now trace the *expectation* of the cumulative sum of the logarithm of withdrawal up to the date of maturity N, for both strategies.

```
MC <- 10000
lR <- matrix(0,N,MC)
lRstar <- matrix(0,N,MC)

for (mc in 1:MC){
  mat <- strategie_conso(r,u,d,N,q)
  R <- c(mat[2:N,1]*conso[1:(N-1)]/(1-conso[1:(N-1)]),
  mat[N,1])
  Rstar<- c(mat[2:N,2]*cstar[1:(N-1)]/(1-cstar[1:(N-1)]),
  mat[N,2])
  lR[,mc] <- cumsum(log(R))
  lRstar[,mc] <- cumsum(log(Rstar))
}

mlR = apply(lR,1,sum)/MC
mlRstar = apply(lRstar,1,sum)/MC

mat <- cbind(mlR, mlRstar)
matplot(1:N, mat, type=c("l"), col=c("blue", "red"))
```

The graphical outputs are given in Figure 8.12. The differences are small; nonetheless, it can be seen that the optimal strategy is indeed that which gives the best performances. The optimal strategy gives a terminal expectation value that is clearly higher than that of the other strategy. Let us verify this by looking at the approximation of the expectation of the cumulative sum of the logarithm of the withdrawal at N for each strategy.

```
lRN=lR[N,]
lRstarN=lRstar[N,]

IC=c(mean(lRN)-1.96*sqrt(var(lRN)/MC), mean(lRN), mean(lRN)
    +1.96*sqrt(var(lRN)/MC))
ICst=c(mean(lRstarN)-1.96*sqrt(var(lRstarN)/MC),
    mean(lRstarN), mean(lRstarN)+1.96*sqrt(var(lRstarN)/MC))
```

We obtain the following values (with a confidence interval at 95%):

– optimal strategy: -258.32 (-258.77; -257.86]);

– other strategy: -276.81 ([-276.97; -276.65]).

The confidence intervals do not intersect; thus, it is possible to state with 95% certainty that the optimal strategy is the best.

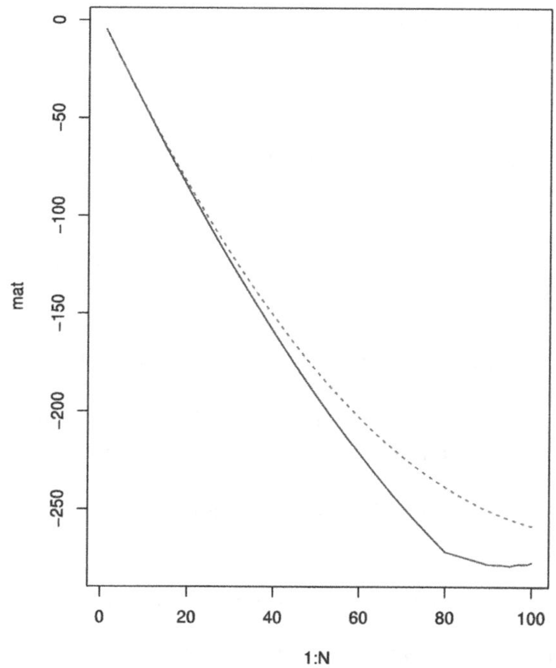

Figure 8.12. *Trajectories of the expectation of the cumulative sum of the logarithm of the withdrawal for the strategies in Exercise 5.6.3, Question 1 (blue) and the optimal strategy (red). For a color version of this figure, see www.iste.co.uk/desaporta/martingales.zip*

8.7. Solutions to exercises in Chapter 6

8.7.1. *Exercise 6.1*

Consider the following market model on $\Omega = \{\omega_1, \omega_2, \omega_3\}$

	$n = 0 \, : \, (S_0^0, S_0^1, S_0^2)$	$n = 1 \, : \, (S_1^0, S_1^1, S_1^2)$
ω_1	$(1, 1, 1)$	$(1.1, 0.66, 1.32)$
ω_2	$(1, 1, 1)$	$(1.1, 1.54, 0.88)$
ω_3	$(1, 1, 1)$	$(1.1, 1.54, 0.88)$

1) Let us calculate the interest rate for the risk-free asset: we directly obtain $r = 10\%$.

The discounted price of the risky assets is

$$\widetilde{S}_1^1(\omega_1) = 0.6, \widetilde{S}_1^1(\omega_2) = \widetilde{S}_1^1(\omega_3) = 1.4,$$

$$\widetilde{S}_1^2(\omega_1) = 1.2, \widetilde{S}_1^2(\omega_2) = \widetilde{S}_1^2(\omega_3) = 0.8.$$

2) Consider the option $X_1(\omega_1) = X_1(\omega_2) = 1$, $X_1(\omega_3) = 1.1$. This option cannot be simulated as all the assets have the same values on the two hazards ω_2 and ω_3, while $X_1(\omega_2) \neq X_1(\omega_3)$.

3) The market is not complete, as X_I cannot be simulated.

4) Let us calculate $\mathbb{E}[\widetilde{S}_1^1 \mid \mathcal{F}_0]$ and $\mathbb{E}[\widetilde{S}_1^2 \mid \mathcal{F}_0]$ for the following two probabilities: $\mathcal{P}_1(\omega_2) = \mathcal{P}_1(\omega_3) = 0.25$, and $\mathcal{P}_2(\omega_2) = 0.2$, $\mathcal{P}_2(\omega_3) = 0.3$.

Since $\mathcal{F}_0 = \{\emptyset, \Omega\}$, the conditional expectation is equal to the expectation. We thus obtain

$$\mathbb{E}_1[\widetilde{S}_1^1] = 0.5 \times 0.6 + 0.25 \times 1.4 + 0.25 \times 1.4 = 1 = \widetilde{S}_0^1$$

$$\mathbb{E}_2[\widetilde{S}_1^1] = 0.5 \times 0.6 + 0.2 \times 1.4 + 0.3 \times 1.4 = 1 = \widetilde{S}_0^1$$

$$\mathbb{E}_1[\widetilde{S}_1^2] = 0.5 \times 1.2 + 0.25 \times 0.8 + 0.25 \times 0.8 = 1 = \widetilde{S}_0^2$$

$$\mathbb{E}_2[\widetilde{S}_1^2] = 0.5 \times 1.2 + 0.2 \times 0.8 + 0.3 \times 0.8 = 1 = \widetilde{S}_0^2$$

It can therefore be seen that under $\mathcal{P}_1$ and $\mathcal{P}_2$, the discounted risky assets are martingales. We found two risk-neutral probabilities, which is another way of demonstrating that the market is not complete.

5) We found two risk-neutral probabilities; thus, the market is viable.

8.7.2. *Exercise 6.2*

Let us show the call-put parity relation in a viable and complete market for call options and put options with the same maturity date N, the same strike price K and the same underlying asset S.

For any $0 \leq n \leq N$, we have

$$\begin{aligned} C_n - P_n &= (1+r)^{n-N} \left(\mathbb{E}^*[(S_N - K)_+ | \mathcal{F}_n] - \mathbb{E}^*[(K - S_N)_+ | \mathcal{F}_n]\right) \\ &= (1+r)^{n-N} \left(\mathbb{E}^*[(S_N - K)_+ - (K - S_N)_+ | \mathcal{F}_n]\right) \\ &= (1+r)^{n-N} \left(\mathbb{E}^*[S_N - K | \mathcal{F}_n]\right) \end{aligned}$$

$$= (1+r)^{n-N} \left(\mathbb{E}^*[S_N|\mathcal{F}_n] - K\right)$$
$$= (1+r)^n \mathbb{E}^*[\widetilde{S}_N|\mathcal{F}_n] - (1+r)^{n-N} K$$
$$= (1+r)^n \widetilde{S}_n - (1+r)^{n-N} K$$
$$= S_n - (1+r)^{n-N} K,$$

where C_n (respectively, P_n) is the value of the call option (respectively, put option) on the date n.

8.7.3. *Exercise 6.3*

Consider the following market model on $\Omega = \{\omega_1, \omega_2, \omega_3, \omega_4\}$:

	$n = 0 : (S_0^0, S_0^1)$	$n = 1 : (S_1^0, S_1^1)$	$n = 2 : (S_2^0, S_2^1)$
ω_1	$(1, 1)$	$(1.1, 0.8)$	$(1.21, 0.64)$
ω_2	$(1, 1)$	$(1.1, 0.8)$	$(1.21, 0.96)$
ω_3	$(1, 1)$	$(1.1, 1.2)$	$(1.21, 0.96)$
ω_4	$(1, 1)$	$(1.1, 1.2)$	$(1.21, 1.44)$

1) Let us show that the market follows a Cox, Ross and Rubinstein model.

We augment the multiplications by $1 + u$ or $1 + d$ at each step with $u = 0.2$ and $d = -0.2$.

2) Let us show that the market is viable and complete.

We have $r = 0.1$ therefore $d < r < u$ and the market is viable and complete. The risk-neutral probability is characterized by $p = (r - d)/(u - d) = (0.1 + 0.2)/(0.2 + 0.2) = 3/4$.

Consider a European call option with strike price $K = 0.9$ and the maturity date $N = 2$.

3) For each ω_i, let us calculate the gain corresponding to this option.

$C_2(\omega_1) = (0.64 - 0.9)_+ = 0$, the option will not be exercised;

$C_2(\omega_2) = C_2(\omega_2) = (0.96 - 0.9)_+ = 0.06$, the option will be exercised;

$C_2(\omega_4) = (1.44 - 0.9)_+ = 0.54$, the option will be exercised.

4) Let us calculate the value of the option at each instant using the backward induction formula.

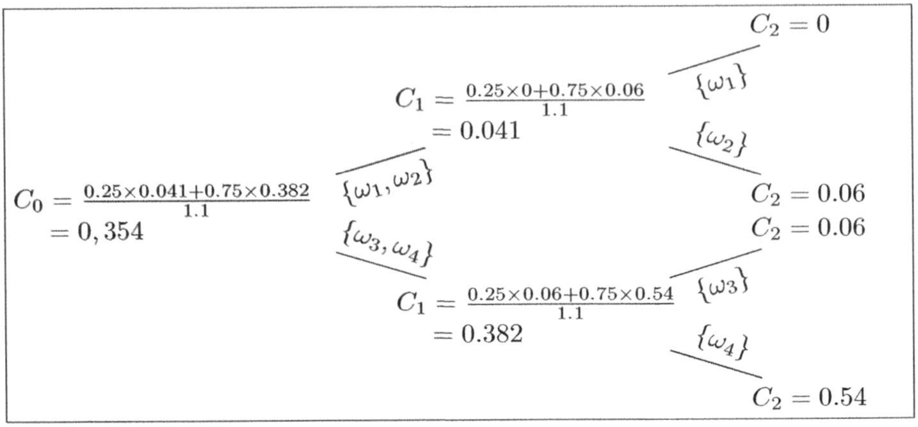

Consider a European put option with the same strike price $K = 0.9$ and maturity date $N = 2$.

5) For each ω_i, let us calculate the gain corresponding to this option.

$P_2(\omega_1) = (0.9 - 0.64)_+ = 0.26$, the option will be exercised;

$P_2(\omega_2) = C_2(\omega_2) = (0.9 - 0.96)_+ = 0$, the option will not be exercised;

$P_2(\omega_4) = (0.9 - 1.44)_+ = 0$, the option will be exercised.

6) Let us calculate the value of the option at each instant using the backward induction formula.

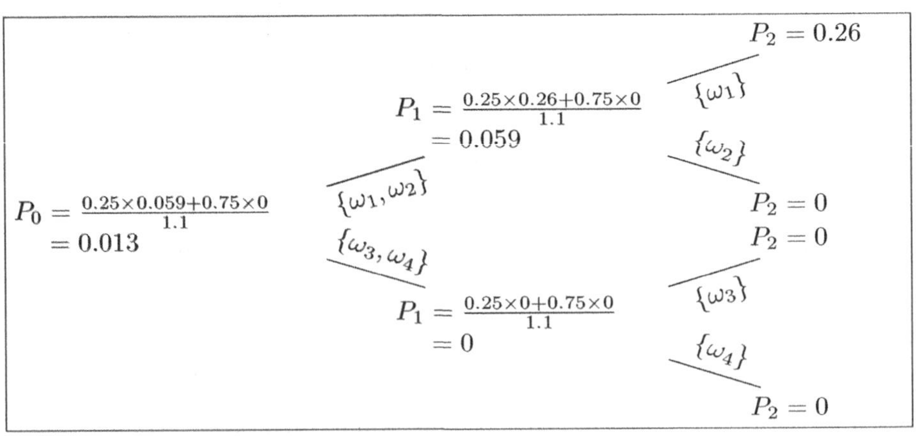

8.7.4. *Exercise 6.4*

Consider the option $X_N = (S_N - K)^2$ in a viable and complete Cox, Ross and Rubinstein market. Let us calculate the value of this option at each instant.

Using the generic formula for the value of an option, we have

$$V_n = (1+r)^{n-N}\mathbb{E}^*[(S_N - K)^2|\mathcal{F}_n]$$

$$= (1+r)^{n-N}\mathbb{E}^*[(S_n T_{n+1}\cdots T_N - K)^2|\mathcal{F}_n].$$

Further, S_n is $\mathcal{F}_n$-measurable, $T_{n+1}\cdots T_N$ is independent of $\mathcal{F}_n$ and takes the values $(1+u)^j(1+d)^{N-n-j}$ with the probability $C^j_{N-n}p^j(1-p)^{N-n-j}$ under $\mathbb{P}^*$. Therefore, we have

$$V_n = (1+r)^{n-N}\sum_{j=0}^{N-n} C^j_{N-n}p^j(1-p)^{N-n-j}(S_N(1+u)^j(1+d)^{N-n-j} - K)^2.$$

8.8. Solution to the practical exercise in Chapter 6 (section 6.6)

1) We wish to write a function R that takes as argument a real value x, an integer n and the parameters r, u, d, K, N and which calculates the function c of the price of a European call option at the point (n, x). In order to do this, we use the explicit form of the value rather than the recursive form. We can note the use of the function pmax which makes it possible to calculate a maximum, coordinate by coordinate, and thus avoid a time loop.

```
fonction_c <- function(x,n,r,u,d,K,N){
  pstar <- -(r-d)/(u-d)
  vecx <- x*((1+u)^(0:(N-n)))*((1+d)^((N-n):0))
  c <-(1+r)^(n-N)*sum(choose(N-n,0:(N-n))*(p^(0:(N-n)))
    *((1-p)^((N-n):0)*pmax(0,vecx)))
  return(c)
}
```

2) On the same graph, let us trace the temporal evolution of the price of the call option C_n of the option and of $(S^1_n - K)_+$. The script that is used is the following, the graphical outputs are given in Figure 8.13.

```
r <- 0.04
u <- 0.1
d <- -0.0
N <- 50
K<- 5
q <- (r-d)/(u-d)
T <- sample(c(1+u,1+d), size=N, replace=TRUE, prob=c(q, 1-q))
S <- c(1,cumprod(T))
C <- NULL
for (i in 0:N){
        C[i+1] <- fonction_c(S[i+1],i,r,u,d,K,N)
}
mat <- rbind(S,C)
matplot(0:N,mat,type=c("l"),col=c("blue","red"))
```

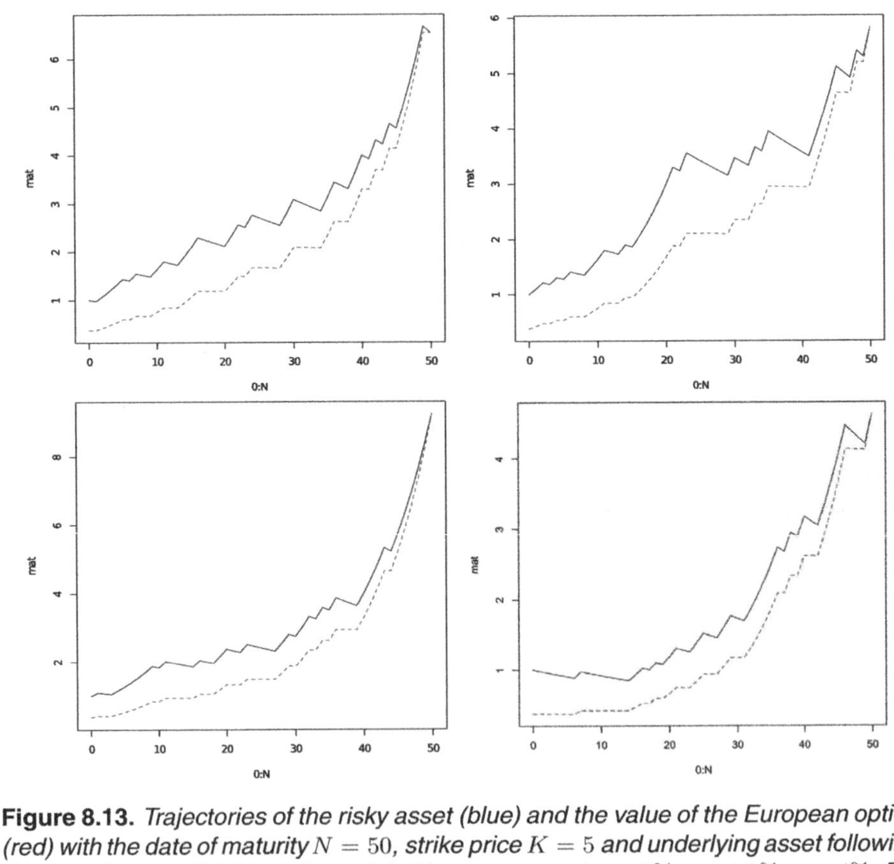

Figure 8.13. *Trajectories of the risky asset (blue) and the value of the European option (red) with the date of maturity $N = 50$, strike price $K = 5$ and underlying asset following the Cox, Ross and Rubinstein model with parameters $d = -2\%$, $u = 10\%$, $r = 4\%$. For a color version of this figure, see www.iste.co.uk/desaporta/martingales.zip*

3) All the simulations give the same value $C_0 = 0.379$. This is normal because C_0 is $\mathcal{F}_0$-measurable, therefore deterministic. This is the value of the option at time 0, i.e. the fair price to pay to acquire the option.

8.9. Solution to exercises in Chapter 7

8.9.1. *Exercise 7.1*

Let $\widetilde{C}_n^a$ be the discounted value at the instant n of an American option with the payoff (Z_n), and let $\widetilde{C}_n^e$ be the discounted value at the instant n of the European option with the same profit Z_N and the same maturity date N.

1) Let us show through backward induction on n that $\widetilde{C}_n^a \geq \widetilde{C}_n^e$ for any $0 \leq n \leq N$.

Initialization: we have $\widetilde{C}_N^a = \widetilde{C}_N^e = \widetilde{Z}_N$; thus, in particular, we do have $\widetilde{C}_N^a \geq \widetilde{C}_N^e$

Hypothesis of recurrence at the rank $n+1$: the property is true for a fixed $n+1$.

Heredity at the rank n: we have

$$\begin{aligned}\widetilde{C}_n^a &= \max\{\widetilde{Z}_n, \mathbb{E}^*[\widetilde{C}_{n+1}^a|\mathcal{F}_n]\}\\ &\geq \mathbb{E}^*[\widetilde{C}_{n+1}^a|\mathcal{F}_n]\\ &\geq \mathbb{E}^*[\widetilde{C}_{n+1}^e|\mathcal{F}_n]\\ &= \mathbb{E}^*\big[\mathbb{E}^*[\widetilde{Z}_N|\mathcal{F}_{n+1}]\big|\mathcal{F}_n\big]\\ &= \mathbb{E}^*[\widetilde{Z}_N|\mathcal{F}_n]\\ &= \widetilde{C}_n^e,\end{aligned}$$

using the hypothesis of recurrence and then the generic formula for the value of a European option.

2) Let us assume that we have $\widetilde{C}_n^e \geq \widetilde{Z}_n$ for any $0 \leq n \leq N$, and let us show that we then have $\widetilde{C}_n^a = \widetilde{C}_n^e$ for any $0 \leq n \leq N$.

If $\widetilde{C}_n^e \geq \widetilde{Z}_n$, then $(\widetilde{C}_n^e)$ is a martingale, therefore a supermartingale greater than $\widetilde{Z}_n$. Since $(\widetilde{C}_n^a)$ is the smallest supermartingale that is greater than $\widetilde{Z}_n$, we have $\widetilde{C}_n^a \leq \widetilde{C}_n^e$ for any $0 \leq n \leq N$, hence the equality.

3) For a call option, it follows

$$\widetilde{C}_n^e = \mathbb{E}^*[(1+r)^{-N}(S_N - K)_+|\mathcal{F}_n] \geq \mathbb{E}^*[\widetilde{S}_N - (1+r)^{-N}K|\mathcal{F}_n].$$

Further, under the risk-neutral probability, the discounted prices are martingales; therefore, we obtain

$$\widetilde{C}_n^e \geq \widetilde{S}_n - (1+r)^{-N}K \geq \widetilde{S}_n - (1+r)^{-n}K.$$

In addition, $\widetilde{C}_n^e \geq 0$, therefore $\widetilde{C}_n^e \geq (\widetilde{S}_n - (1+r)^{-n}K)_+ = \widetilde{Z}_n$. According to the previous question, we can derive from this that $\widetilde{C}_n^e = \widetilde{C}_n^a$ for any n.

It can be noted that this working cannot be applied to put options.

8.9.2. *Exercise 7.2*

Consider an American call option with maturity date N and strike price K in a market that follows the Cox, Ross and Rubinstein model. C_n denotes the value of the option at the time n. We wish to show through backward induction on n that $C_n = g(n, S_n^1)$ with

$$\begin{cases} g(N,x) = (x-K)_+, \\ g(n,x) = \max\left\{(x-K)_+;\ \frac{pg\big(n+1,x(1+u)\big)+(1-p)g\big(n+1,x(1+d)\big)}{1+r}\right\}. \end{cases}$$

Initialization: For $n = N$, we have $C_N = Z_N = (S_N - K)_+$.
Heredity: Let us assume that the formula is true at the rank $n+1$ for a fixed n. We then have

$$C_n = \max\{Z_n; \frac{1}{1+r}\mathbb{E}^*[C_{n+1}|\mathcal{F}_n]\}.$$

Further, we have

$$\begin{aligned} \mathbb{E}^*[[C_{n+1}|\mathcal{F}_n]\} &= \mathbb{E}^*[g(n+1, S_{n+1}^1|\mathcal{F}_n]\} \\ &= \mathbb{E}^*[g(n+1, S_n^1 T_{n+1}|\mathcal{F}_n]\} \\ &= (1-p)g(n+1, S_n^1(1+d)) + pg(n+1, S_n^1(1+u)), \end{aligned}$$

using the independence of T_n under $\mathbb{P}^*$, hence the result.

8.9.3. *Exercise 7.3*

Consider a financial market following the Cox, Ross and Rubinstein model with parameters $r = 0$, $d = -0.1$ and $u = 0.1$ on $\Omega = \{\omega_1, \dots, \omega_8\}$. The risky asset takes the following values:

	$n=0$	$n=1$	$n=2$	$n=3$
ω_1	1	1.1	1.21	1.331
ω_2	1	1.1	1.21	1.089
ω_3	1	1.1	0.99	1.089
ω_4	1	1.1	0.99	0.891
ω_5	1	0.9	0.99	1.089
ω_6	1	0.9	0.99	0.891
ω_7	1	0.9	0.81	0.891
ω_8	1	0.9	0.81	0.729

1) We do have $d < r < u$; therefore, the market is viable and complete. Further, $p = \frac{r-d}{u-d} = 0.1/0.2 = 1/2$.

We consider an American call option on this market with strike price $K = 0.9$ and maturity date $N = 3$.

2) For every instant and every ω_i, let us calculate the payoff from exercising the option. It is known that $Z_n = (S_n - K)_+$, which gives the following tree.

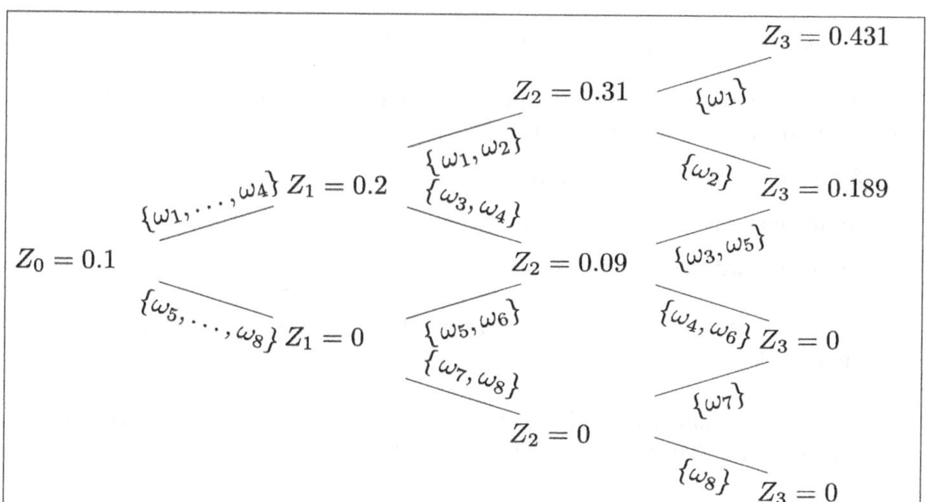

3) For each instant and each ω_i, let us calculate the value of the option. We use the induction formula for $r = 0$ and $p = 1/2$.

$$\begin{cases} f(N,x) = (x-K)_+ \\ f(n,x) = \max\{(x-K)_+; (1-p)f(n+1, x(1+d)) + p)f(n+1, x(1+u))\}. \end{cases}$$

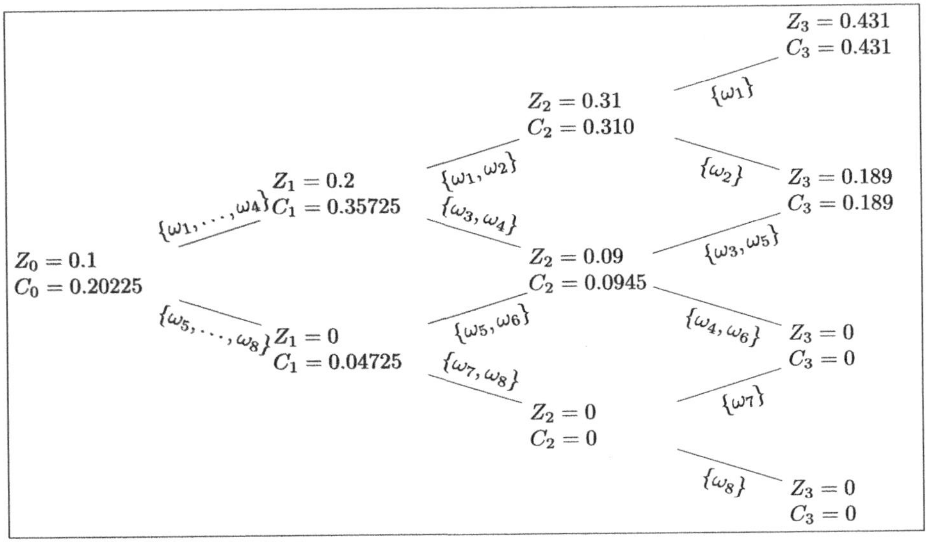

4) For each ω_i, let us given an optimal date for exercising the option.

For example, we can choose the first instant where $Z_n = C_n$ (the stopping time $\overline{T}$ is not easy to calculate, in practice!):

– for the hazards ω_1, ω_2, we have $n = 2$;

– for the hazards $\omega_3, \omega_4, \omega_4, \omega_6$, we have $n = 3$;

– for the hazards ω_7, ω_8, we have $n = 2$.

Consider an American put option on this market with strike price $K = 0.9$ and maturity date $N = 3$.

5) For each instant and each ω_i, let us calculate the profit from exercising the option. It is known that $Z_n = (K - S_n)_+$, which gives the following tree.

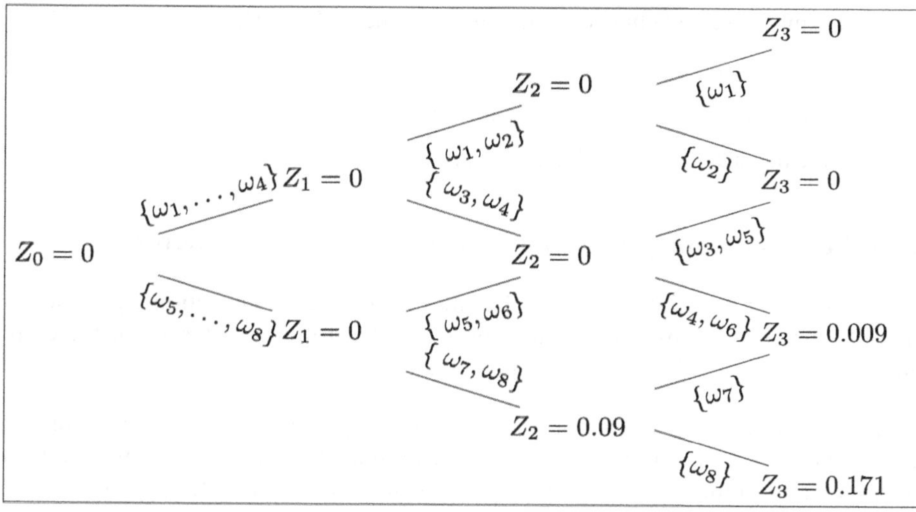

6) For each instant and each ω_i, let us calculate the value of the option. We use the induction formula for $r = 0$ and $p = 1/2$.

$$\begin{cases} f(N, x) = (K - x)_+ \\ f(n, x) = \max\{(K - x)_+; (1-p)f(n+1, x(1+d)) + p)f(n+1, x(1+u))\}. \end{cases}$$

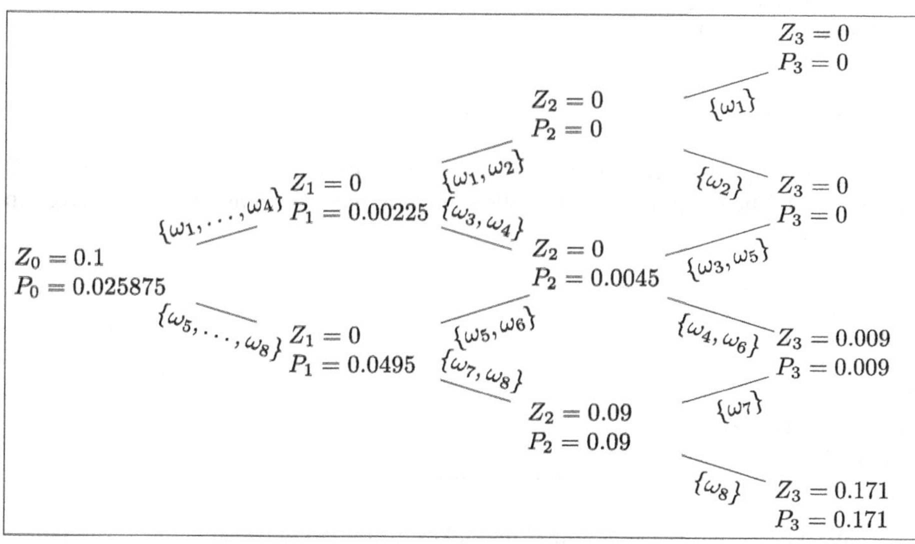

7) For each ω_i, let us given an optimal date for exercising the option.

For example, we can choose the first instant where $Z_n = P_n$:

– for the hazards ω_1, ω_2, we have $n = 2$;

– for the hazards $\omega_3, \omega_4, \omega_4, \omega_6$, we have $n = 3$;

– for the hazards ω_7, ω_8 we have $n = 2$.

8.10. Solution to the practical exercise in Chapter 7 (section 7.6)

We will now illustrate how to calculate the American put options. In order to simplify the simulations, we will assume in all that follows that $r = 0$, therefore the processes are equal to discounted processes.

1) We wish to write a function R that takes as argument the parameters u and d, a strike price K and a horizon N and returns a trajectory $(Z_n)_{0 \leq n \leq N}$ for the gain corresponding to the exercising of the American option with underlying asset (S_n) following the Cox, Ross and Rubinstein model, with strike price K and maturity date N. In order to do this, we directly use the formula $Z_n = (K - S_n^1)_+$.

```
trajZ <- function(u,d,K,N){
  pstar <- -d/(u-d)
  X<- sample(c(1+u,1+d), size=N, replace=TRUE,
        prob=c(pstar, 1-pstar))
  S<-c(1,cumprod(X))
  Z<-pmax(K-S,0)
  return(Z)
}
```

2) Let us graphically represent some trajectories for $u = 10\%$, $d = -5\%$, $K = 1.2$ and $N = 50$. The script used is the following, and the graphical outputs are given in Figure 8.14.

```
u <- 0.1
d <- -0.05
N <- 50
K<- 1.2
traj <- trajZ(u,d,N)
plot(0:N, traj, type="b")
```

3) We wish to write a function R that takes as argument a real value x, an integer n and the parameters u, d, K, N and which calculates the function f, for the price of an American put option at the point (n, x). In order to do this, we use the induction formula on this function, which gives us a recursive function.

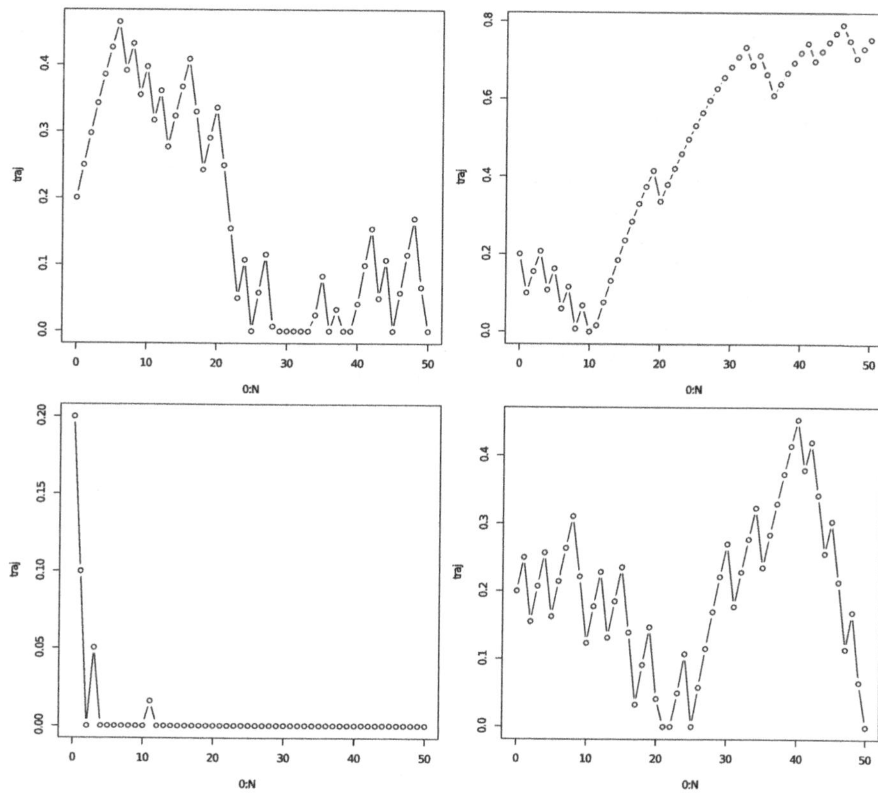

Figure 8.14. *Trajectories of the payoff for the American option with maturity date $N = 50$, with strike price $K = 1.2$ and an underlying asset following the Cox, Ross and Rubinstein model with parameters $d = -5\%$, $u = 10\%$ and $r = 0$*

```
recf <- function(x,n,u,d,K,N){
  pstar <- -d/(u-d)
  if (n==N){
    f <- max(K-x,0)
  }
  else{
    f <-max(max(K-x,0),
            recf(x*(1+u),n+1,u,d,K,N)*pstar+recf(x*(1+d),
               n+1,u,d,K,N)*(1-pstar))
  }
  return(f)
}
```

4) Let us now write a function R that takes as argument the parameters u and d, a strike price K and a horizon N and returns a trajectory $(U_n, Z_n)_{0 \leq n \leq N}$ for the gain corresponding to the exercise and value of the American option with underlying asset

(S_n) following the Cox, Ross and Rubinstein model, with strike price K and maturity date N. We use the function defined in the previous question.

```
trajU <- function(u,d,K,N){
        pstar <- -d/(u-d)
        X <- sample(c(1+u,1+d), size=N, replace=TRUE,
          prob=c(pstar, 1-pstar))
        S <- c(1,cumprod(X))
        Z <- pmax(K-S,0)
        U <- NULL
        U[N+1] <- Z[N+1]
        for (i in seq(N,1,by=-1)){
                U[i] <- recf(S[i],i-1,u,d,K,N)
        }
        return(cbind(U,Z))
}
```

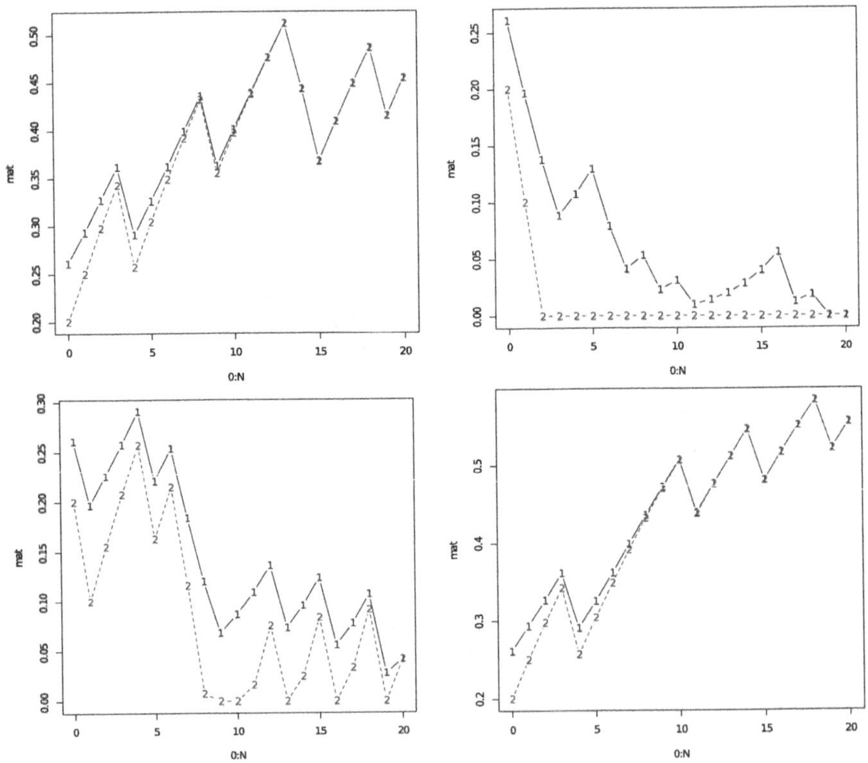

Figure 8.15. *Trajectories for the payoff (red) and for the value (blue) of an American option with maturity date $N = 50$, strike price $K = 1.2$ and an underlying asset following the Cox, Ross and Rubinstein model with parameters $d = -5\%$, $u = 10\%$ and $r = 0$. For a color version of this figure, see www.iste.co.uk/desaporta/martingales.zip*

5) Let us graphically represent some trajectories of (U, Z) for $u = 10\%$, $d = -5\%$, $K = 1.2$ and $N = 20$. The script used is as follows, and the graphical outputs are given in Figure 8.15.

```
u <- 0.1
d <- -0.05
N <- 50
K <- 1.2
mat <- trajU(u,d,K,N)
matplot(0:N,mat,type=c("b"),col=c("blue","red"))
```

```
> mat
            U         Z
 [1,] 0.2609258 0.2000000
 [2,] 0.2931552 0.2500000
 [3,] 0.3268698 0.2975000
 [4,] 0.3616882 0.3426250
 [5,] 0.2906972 0.2568875
 [6,] 0.3257350 0.3040431
 [7,] 0.3618845 0.3488410
 [8,] 0.3986309 0.3913989
 [9,] 0.4354415 0.4318290
[10,] 0.3627099 0.3550119
[11,] 0.4008504 0.3972613
[12,] 0.4387844 0.4373982
[13,] 0.4759264 0.4755283
[14,] 0.5118165 0.5117519
[15,] 0.4431208 0.4429271
[16,] 0.3678009 0.3672198
[17,] 0.4088588 0.4088588
[18,] 0.4484159 0.4484159
[19,] 0.4859951 0.4859951
[20,] 0.4145946 0.4145946
[21,] 0.4538648 0.4538648
```

$T_0 = 16$

```
> mat
             U   Z
 [1,] 0.26092582 0.2
 [2,] 0.19646706 0.1
 [3,] 0.13794928 0.0
 [4,] 0.08842394 0.0
 [5,] 0.10752684 0.0
 [6,] 0.12967037 0.0
 [7,] 0.07902539 0.0
 [8,] 0.04121708 0.0
 [9,] 0.05326357 0.0
[10,] 0.02320669 0.0
[11,] 0.03127156 0.0
[12,] 0.01007319 0.0
[13,] 0.01429644 0.0
[14,] 0.02022465 0.0
[15,] 0.02850695 0.0
[16,] 0.04001540 0.0
[17,] 0.05590555 0.0
[18,] 0.01235264 0.0
[19,] 0.01852896 0.0
[20,] 0.00000000 0.0
[21,] 0.00000000 0.0
```

$T_0 = 19$

Table 8.5. *Trajectories of the payoff (Z) and of the value (U) of the American option with maturity date $N = 50$, strike price $K = 1.2$ and underlying asset following the Cox, Ross and Rubinstein model with parameters $d = -5\%$, $u = 10\%$ and $r = 0$*

It can be noted that the number of calls to the function `recf` is very high, which slows down the code significantly. Further, we always recalculate the same value (e.g. see the trees in Exercise 7.3). It would, therefore, be more digitally efficient to store the values that have already been calculated and to reuse them.

```
> mat
               U           Z
 [1,] 0.26092582 0.200000000
 [2,] 0.19646706 0.100000000
 [3,] 0.22572595 0.155000000
 [4,] 0.25723295 0.207250000
 [5,] 0.29069717 0.256887500
 [6,] 0.22062158 0.162576250
 [7,] 0.25343594 0.214447437
 [8,] 0.18352452 0.115892181
 [9,] 0.12026252 0.007481399
[10,] 0.06829201 0.000000000
[11,] 0.08680224 0.000000000
[12,] 0.10926798 0.016127159
[13,] 0.13607302 0.075320801
[14,] 0.07337453 0.000000000
[15,] 0.09580832 0.024710237
[16,] 0.12370478 0.083474725
[17,] 0.05590555 0.000000000
[18,] 0.07768200 0.033231088
[19,] 0.10725852 0.091569534
[20,] 0.02779344 0.000000000
[21,] 0.04169016 0.041690163
```

$T_0 = 20$

```
> mat
              U         Z
 [1,] 0.2609258 0.2000000
 [2,] 0.2931552 0.2500000
 [3,] 0.3268698 0.2975000
 [4,] 0.3616882 0.3426250
 [5,] 0.2906972 0.2568875
 [6,] 0.3257350 0.3040431
 [7,] 0.3618845 0.3488410
 [8,] 0.3986309 0.3913989
 [9,] 0.4354415 0.4318290
[10,] 0.4718073 0.4702375
[11,] 0.5072858 0.5067256
[12,] 0.4387844 0.4373982
[13,] 0.4759264 0.4755283
[14,] 0.5118165 0.5117519
[15,] 0.5461643 0.5461643
[16,] 0.4807807 0.4807807
[17,] 0.5167417 0.5167417
[18,] 0.5509046 0.5509046
[19,] 0.5833594 0.5833594
[20,] 0.5216953 0.5216953
[21,] 0.5556105 0.5556105
```

$T_0 = 14$

Table 8.6. *Trajectories of the payoff (Z) and of the value (U) of the American option with maturity date $N = 50$, strike price $K = 1.2$ and underlying asset following the Cox, Ross and Rubinstein model with parameters $d = -5\%$, $u = 10\%$ and $r = 0$*

6) For each of our graphs, determine the value for the optimal stopping time T_0. In order to do this, we display the values of the sequences U and Z to find the first instance where they coincide. We obtain the results in Tables 8.5 and 8.6.

7) All the simulations give the same value $U_0 = 0.261$. This is normal because U_0 is $\mathcal{F}_0$-measurable and thus deterministic. This is the value of the option at the time 0, i.e. the fair price to be paid to acquire the option.

References

[BAL 00] BALDI P., MAZLIAK L., PRIOURET P., *Martingales et Chaînes de Markov*, Hermann, Paris, 2000.

[BAR 07] BARBE P., LEDOUX M., *Probabilité*, EDP Science, Les Ulis, 2007.

[BIL 12] BILLINGSLEY P., *Probability and Measure*, John Wiley & Sons, Inc., Hoboken, NJ, 2012.

[BLA 73] BLACK F., SCHOLES M., "The pricing of options and corporate liabilities", *Journal of Political Economy*, vol. 81, no. 3, pp. 637–654, 1973.

[BRÉ 99] BRÉMAUD P., *Markov Chains: Gibbs Fields, Monte Carlo Simulation, and Queues*, Springer-Verlag, New York, 1999.

[CHU 01] CHUNG K.L., *A Course in Probability Theory*, 3rd edition, Academic Press, Inc., San Diego, CA, 2001.

[DAC 82] DACUNHA-CASTELLE D., DUFLO M., *Probabilités et statistiques : problèmes à temps fixe*, Volume 1, Masson, Paris, 1982.

[DAC 83] DACUNHA-CASTELLE D., DUFLO M., *Probabilités et statistiques : problèmes à temps mobile*, Volume 2, Masson, Paris, 1983.

[DUF 97] DUFLO M., *Random Iterative Models*, Springer-Verlag, Berlin, 1997.

[DUR 10] DURRETT R., *Probability: Theory and Examples*, 4th edition, Cambridge University Press, Cambridge, 2010.

[FEL 68] FELLER W., *An Introduction to Probability Theory and Its Applications*, Volume I, 3rd edition, John Wiley & Sons Inc., New York, 1968.

[FOA 03] FOATA D., FUCHS A., *Calcul des probabilités*, 2nd edition, Dunod, Paris, 2003.

[HUL 18] HULL J.C., *Options, Futures, and Other Derivatives*, 10th edition, Pearson, New York, 2018.

[KAL 02] KALLENBERG O., *Foundations of Modern Probability*, 2nd edition, Springer-Verlag, New York, 2002.

[LAM 97] LAMBERTON D., LAPEYRE B., *Introduction au calcul stochastique appliqué à la finance*, 2nd edition, Ellipses Édition Marketing, Paris, 1997.

[MER 69] MERTON R.C., "Lifetime portfolio selection under uncertainty: The continuous-time case", *The Review of Economics and Statistics*, vol. 51, no. 3, pp. 247–257, 1969.

[NOR 98] NORRIS J.R., *Markov Chains*, Cambridge University Press, Cambridge, 1998.

[OUV 08] OUVRARD J.-Y., *Probabilités*, Volume 1, Cassini, Paris, 2008.

[OUV 09] OUVRARD J.-Y., *Probabilités*, Volume 2, Cassini, Paris, 2009.

[SHI 00] SHIRYAEV A., *Probability*, Springer, New York, 2000.

[SHR 03] SHREVE S., *Stochastic Calculus for Finance I: The Binomial Asset Pricing Model*, Springer, New York, 2003.

[SHR 04] SHREVE S., *Stochastic Calculus for Finance II: Continuous-Time Models*, Springer, New York, 2004.

[VIN 04] VINCENZO CAPASSO D.B., *An Introduction to Continuous Time Stochastic Processes: Theory, Models, and Applications to Finance, Biology, and Medicine*, Birkhäuser, Boston, MA, 2004.

[WIL 91] WILLIAMS D., *Probability with Martingales*, Cambridge University Press, Cambridge, 1991.

Index

A, B

adapted, 18
arbitrage opportunity, 85
Black and Scholes, 86

C

Cauchy-Schwarz, 33
Chain rule, 35
conditional
  expectation, 26
  probability, 21
convergence
  almost sure, 16
  dominated, 32
  in L^2, 15
  in L^p, 15
  monotone, 16, 30
  strong law of large numbers, 16
Cox, Ross and Rubinstein, 86

D

discounted prices, 82
discrete random variables, 8
distribution
  conditional, 22
  finite-dimensional, 17
  joint distribution, 14
  marginal distribution, 14
  of a random variable, 8
  probability, 5
Doob
  decomposition theorem, 67
  stopping theorem, 73

E, F

events, 2
expectation, 10
Fatou, 31
filtration natural, 18

G, H

generating system, 3
Hilbert, 37
hitting time, 70

I, J

independence, 6
independent increments, 65
integrable, 10
investment strategy, 82
  hedging strategy, 97
Jensen, 34

M

market
  complete, 96
  financial, 82
  viable, 85

martingale, 63
  transform, 66

O

optimal stopping, 109
  time, 110
option
  American, 107
  call, 95
  Call-Put parity relation, 105
  European, 95
  exercise, 114
  hedging, 114
    strategy, 97
  maturity, 95
  options exotic, 96
  payoff, 96
  put, 96
  simulable, 96
  strike price, 95
  underlying, 95
  value, 97
orthogonal projection, 37

P

paths, 49
payoff, 107
predictable, 18
probability space, 2

R

random walk, 46
  trajectory of, 47
real random vector, 14
reflection, 50
risk-
  free interest rate, 82
  neutral probability, 85
risky assets
  price of, 82

S

σ-algebra, 2
  generated by a random variable, 12
  of events observable up until time, 70
Snell envelope, 110
stochastic process, 16
stopped martingale, 71
stopping time, 69
strategy
  admissible, 84
  arbitrage, 85
  self-financed, 83
submartingale, 63
supermartingale, 63

T, U, V

terminal variable, 73
usual distributions
  Bernoulli distribution, 9
  binomial distribution, 9
  hypergeometric distribution, 10
  Poisson distribution, 10
  uniform distribution, 9
variance, 12

Other titles from

iSTE

in

Mathematics and Statistics

2021

KOROLIOUK Dmitri, SAMOILENKO Igor
Random Evolutionary Systems: Asymptotic Properties and Large Deviations

MOKLYACHUK Mikhail
Convex Optimization: Introductory Course

POGORUI Anatoliy, SWISHCHUK Anatoliy, RODRÍGUEZ-DAGNINO Ramón M.
Random Motions in Markov and Semi-Markov Random Environments 1: Homogeneous Random Motions and their Applications
Random Motions in Markov and Semi-Markov Random Environments 2: High-dimensional Random Motions and Financial Applications

PROVENZI Edoardo
From Euclidean to Hilbert Spaces: Introduction to Functional Analysis and its Applications

2020

BARBU Vlad Stefan, VERGNE Nicolas
Statistical Topics and Stochastic Models for Dependent Data with Applications

CHABANYUK Yaroslav, NIKITIN Anatolii, KHIMKA Uliana
Asymptotic Analyses for Complex Evolutionary Systems with Markov and Semi-Markov Switching Using Approximation Schemes

KOROLIOUK Dmitri
Dynamics of Statistical Experiments

MANOU-ABI Solym Mawaki, DABO-NIANG Sophie, SALONE Jean-Jacques
Mathematical Modeling of Random and Deterministic Phenomena

2019

BANNA Oksana, MISHURA Yuliya, RALCHENKO Kostiantyn, SHKLYAR Sergiy
Fractional Brownian Motion: Approximations and Projections

GANA Kamel, BROC Guillaume
Structural Equation Modeling with lavaan

KUKUSH Alexander
Gaussian Measures in Hilbert Space: Construction and Properties

LUZ Maksym, MOKLYACHUK Mikhail
Estimation of Stochastic Processes with Stationary Increments and Cointegrated Sequences

MICHELITSCH Thomas, PÉREZ RIASCOS Alejandro, COLLET Bernard, NOWAKOWSKI Andrzej, NICOLLEAU Franck
Fractional Dynamics on Networks and Lattices

VOTSI Irene, LIMNIOS Nikolaos, PAPADIMITRIOU Eleftheria, TSAKLIDIS George
Earthquake Statistical Analysis through Multi-state Modeling (Statistical Methods for Earthquakes Set – Volume 2)

2018

AZAÏS Romain, BOUGUET Florian
Statistical Inference for Piecewise-deterministic Markov Processes

IBRAHIMI Mohammed
Mergers & Acquisitions: Theory, Strategy, Finance

PARROCHIA Daniel
Mathematics and Philosophy

2017

CARONI Chysseis
First Hitting Time Regression Models: Lifetime Data Analysis Based on Underlying Stochastic Processes
(Mathematical Models and Methods in Reliability Set – Volume 4)

CELANT Giorgio, BRONIATOWSKI Michel
Interpolation and Extrapolation Optimal Designs 2: Finite Dimensional General Models

CONSOLE Rodolfo, MURRU Maura, FALCONE Giuseppe
Earthquake Occurrence: Short- and Long-term Models and their Validation
(Statistical Methods for Earthquakes Set – Volume 1)

D'AMICO Guglielmo, DI BIASE Giuseppe, JANSSEN Jacques, MANCA Raimondo
Semi-Markov Migration Models for Credit Risk
(Stochastic Models for Insurance Set – Volume 1)

GONZÁLEZ VELASCO Miguel, del PUERTO GARCÍA Inés, YANEV George P.
Controlled Branching Processes
(Branching Processes, Branching Random Walks and Branching Particle Fields Set – Volume 2)

HARLAMOV Boris
Stochastic Analysis of Risk and Management
(Stochastic Models in Survival Analysis and Reliability Set – Volume 2)

KERSTING Götz, VATUTIN Vladimir
Discrete Time Branching Processes in Random Environment
(Branching Processes, Branching Random Walks and Branching Particle Fields Set – Volume 1)

MISHURA YULIYA, SHEVCHENKO Georgiy
Theory and Statistical Applications of Stochastic Processes

NIKULIN Mikhail, CHIMITOVA Ekaterina
Chi-squared Goodness-of-fit Tests for Censored Data
(Stochastic Models in Survival Analysis and Reliability Set – Volume 3)

SIMON Jacques
Banach, Fréchet, Hilbert and Neumann Spaces
(Analysis for PDEs Set – Volume 1)

2016

CELANT Giorgio, BRONIATOWSKI Michel
Interpolation and Extrapolation Optimal Designs 1: Polynomial Regression and Approximation Theory

CHIASSERINI Carla Fabiana, GRIBAUDO Marco, MANINI Daniele
Analytical Modeling of Wireless Communication Systems
(Stochastic Models in Computer Science and Telecommunication Networks Set – Volume 1)

GOUDON Thierry
Mathematics for Modeling and Scientific Computing

KAHLE Waltraud, MERCIER Sophie, PAROISSIN Christian
Degradation Processes in Reliability
(Mathematial Models and Methods in Reliability Set – Volume 3)

KERN Michel
Numerical Methods for Inverse Problems

RYKOV Vladimir
Reliability of Engineering Systems and Technological Risks
(Stochastic Models in Survival Analysis and Reliability Set – Volume 1)

2015

DE SAPORTA Benoîte, DUFOUR François, ZHANG Huilong
Numerical Methods for Simulation and Optimization of Piecewise Deterministic Markov Processes

DEVOLDER Pierre, JANSSEN Jacques, MANCA Raimondo
Basic Stochastic Processes

LE GAT Yves
Recurrent Event Modeling Based on the Yule Process
(Mathematical Models and Methods in Reliability Set – Volume 2)

2014

COOKE Roger M., NIEBOER Daan, MISIEWICZ Jolanta
Fat-tailed Distributions: Data, Diagnostics and Dependence
(Mathematical Models and Methods in Reliability Set – Volume 1)

MACKEVIČIUS Vigirdas
Integral and Measure: From Rather Simple to Rather Complex

PASCHOS Vangelis Th
Combinatorial Optimization – 3-volume series – 2nd edition
Concepts of Combinatorial Optimization / Concepts and Fundamentals – volume 1
Paradigms of Combinatorial Optimization – volume 2
Applications of Combinatorial Optimization – volume 3

2013

COUALLIER Vincent, GERVILLE-RÉACHE Léo, HUBER Catherine, LIMNIOS Nikolaos, MESBAH Mounir
Statistical Models and Methods for Reliability and Survival Analysis

JANSSEN Jacques, MANCA Oronzio, MANCA Raimondo
Applied Diffusion Processes from Engineering to Finance

SERICOLA Bruno
Markov Chains: Theory, Algorithms and Applications

2012

BOSQ Denis
Mathematical Statistics and Stochastic Processes

CHRISTENSEN Karl Bang, KREINER Svend, MESBAH Mounir
Rasch Models in Health

DEVOLDER Pierre, JANSSEN Jacques, MANCA Raimondo
Stochastic Methods for Pension Funds

2011

MACKEVIČIUS Vigirdas
Introduction to Stochastic Analysis: Integrals and Differential Equations

MAHJOUB Ridha

Recent Progress in Combinatorial Optimization – ISCO2010

RAYNAUD Hervé, ARROW Kenneth
Managerial Logic

2010

BAGDONAVIČIUS Vilijandas, KRUOPIS Julius, NIKULIN Mikhail
Nonparametric Tests for Censored Data

BAGDONAVIČIUS Vilijandas, KRUOPIS Julius, NIKULIN Mikhail
Nonparametric Tests for Complete Data

IOSIFESCU Marius *et al.*
Introduction to Stochastic Models

VASSILIOU PCG
Discrete-time Asset Pricing Models in Applied Stochastic Finance

2008

ANISIMOV Vladimir
Switching Processes in Queuing Models

FICHE Georges, HÉBUTERNE Gérard
Mathematics for Engineers

HUBER Catherine, LIMNIOS Nikolaos *et al.*
Mathematical Methods in Survival Analysis, Reliability and Quality of Life

JANSSEN Jacques, MANCA Raimondo, VOLPE Ernesto
Mathematical Finance

2007

HARLAMOV Boris
Continuous Semi-Markov Processes

2006

CLERC Maurice
Particle Swarm Optimization

Printed and bound by CPI Group (UK) Ltd, Croydon, CR0 4YY

06/05/2026

14872631-0001